AN INTRODUCTION TO
WAR
THEORY

D1343178

Can a battle ever be decisive? Stalingrad, perhaps, or Trafalgar? They are not easy to identify. One has only to ask what difference it would have made to the outcome of the Second World War if the Germans had held the ruins as securely as they had Paris. Was the apparent maritime Pax Britannica that followed Trafalgar the result of the victory, or was the battle merely a memorable historical marker in the flow of events? In AD 451, Attila the Hun's vast army was met at Aurelianum in what is now northeast France by the Roman army of Flavius Aetius. After the tremendous slaughter, the Romans left the field; but the Huns were exhausted and turned for home. Historian John Julius Norwich considered that 'the whole fate of western civilisation hung in the balance. Had the Hunnish leader not been halted … both Gaul and Italy would have been reduced to spiritual and cultural deserts.' Few people can name the Battle of the Catalaunian Plains. (By Victor Ambrus from *Battlefield Panoramas*, The History Press)

AN INTRODUCTION TO
WAR
THEORY

DR CHRIS BROWN

About the Author

Chris Brown has led courses on medieval history for St Andrews and Edinburgh universities. He has also appeared as consultant on numerous documentaries and conducts heritage tours at the battlefield sites. He is the author of *Battle Story: Arnhem* and *Battle Story: Kohima*, *The Scottish Wars of Independence*, *Robert the Bruce: A Life Chronicled* and *Bannockburn 1314: The Battle 700 Years On*. He lives on the Shetland Islands.

Cover illustrations: Two blue-eyed military strategists. A contemplative Hitler and Major-General Carl von Clausewitz, painted by Karl Wilhelm Wach (1830). Clausewitz was the first to use the term *Nebel des Krieges*– the fog of war. (Both Wikimedia Commons); *Above*: Assyrian archers portrayed at the palace of Balawat, Iraq. The palace was built in the ninth century BC. (Istanbul Archaeological Museum)

First published 2014

The History Press
The Mill, Brimscombe Port
Stroud, Gloucestershire, GL5 2QG
www.thehistorypress.co.uk

© Dr Chris Brown, 2014

The right of Dr Chris Brown to be identified as the Author
of this work has been asserted in accordance with the
Copyright, Designs and Patents Act 1988.

All rights reserved. No part of this book may be reprinted
or reproduced or utilised in any form or by any electronic,
mechanical or other means, now known or hereafter invented,
including photocopying and recording, or in any information
storage or retrieval system, without the permission in writing
from the Publishers.

British Library Cataloguing in Publication Data.
A catalogue record for this book is available from the British Library.

ISBN 978 0 7509 5972 8

Typeset in 10.5/14pt Bembo by Thomas Bohm, User design
Printed in Great Britain

Contents

Introduction and Acknowledgements

No single volume could possibly provide a complete guide to the theory of war any more than one book could give any more than the most cursory overview of a history of war or the practice of war. As Lieutenant Colonel Ferdinand Foch wrote in *Principles of War*, 'Everything in war is linked together, is mutually interdependent, mutually interpenetrating.'

The purpose of this book is to provide a starting point for more extensive and deeper study or to give some basic groundwork for anyone who has an interest in the principles of how wars come about and how they are pursued. It is more akin to a volume on basic mechanics rather than a workshop manual for a Ford or a Subaru. It will not teach the reader how to be a general; that takes a great deal of training and assiduous study … not to mention twenty years of service in the army.

In part, this book was prompted by teaching a military history class for Edinburgh University's OLL department but it is also the result of finding myself in a conference room full of professional history scholars. It transpired that all of them had written essays, papers, articles and even books on wars, but not one had ever read a war theory textbook. I found that a little worrying as well as bewildering.

The pre-eminent military historian Sir Michael Howard, having lectured young army officers at Sandhurst on the Italian Campaign of 1942, was asked a single question by an impatient captain: OK, but what were its *lessons*? He found it a tricky one to answer! That, too, is part of the aim of this book, to explore what, if anything, we can learn from history.

All books are personal – we write them in the light of everything we ever read or heard or deduced in connection with the subject. For that reason the actions referred to in this book have been chosen from the conflicts that interest me. They are no more or less significant than any number of other battles and campaigns and there are others which might have made more appropriate examples, or they might have been more interesting to a wider range of readers, but these are the ones with which I am familiar.

As ever I am indebted to my wife Pat, who has borne the process with immense patience and fortitude. I am also indebted to Sean Anderson, Mike McEwan, Colonel Mark Boden (T.D.) John Neicho and Jimmy Moncrieffe for listening to me going on and on about war theory – among other things – and to Professor Donald Vandergriffe, the late Brigadier M.R.J. Hope Thompson, RSM J. Deeley, Majors D. MacDonald and G.M.A. Athey and my editor Shaun Barrington for various observations and suggestions – especially with regard to the illustrations.

PART 1

HOW DOES WAR HAPPEN?

1. Is War Inevitable?

The short answer is no, but that is very clearly not a lesson that has been learned, so perhaps we should consider why not. It is not the case that the horror of the entire process cannot be understood; it can be, but there are people who do not want to understand it. It is far too easy to simply blame politicians, though almost all warfare has been waged for political purposes – political in the very widest sense, including religious and ideological beliefs.

It is also far too simple to assert that war can only be waged if the people accept the government's instructions. Very often they have no choice in the matter. People who stand their ground and refuse to do military service because they do not believe in the cause espoused or because they do not believe in war under any circumstances are making a courageous choice, but they are made of sterner stuff than most of us. In the past, they have risked hostility, imprisonment, violence, even death rather than serve in the army. Formalised conscription by the state is increasingly rare for a number of reasons, not least the economic burden and the increasingly complicated demands on the infantry soldier, which must be mastered if he is to be a really effective combatant. What we might call 'informal conscription' is very much more common than we might at first imagine and can been seen at its worst in the practice of kidnapping boys to serve in battle. The powers that kidnap these child soldiers are not especially interested in turning them into competent troops for battle so much as constructing armed bands that can terrorise civilian communities.

Experience should have shown us that a government does not have to have the support of the people in order to go to war. Where there is a small professional regular army a government simply makes a decision to project its power and then does so, to the extent that it can, in pursuit of whatever objective (usually political) it has identified. In the United Kingdom, public sentiment was largely opposed to war in Iraq and in Afghanistan, and for good reason. Neither country was a threat to the UK, there were no coherent strategic objectives, and, with the best will in the world, the United Kingdom does not have what might be called an outstanding track record in conflicts in either region. Both of those conflicts were widely seen as a pointless waste of lives and money and as being pursued for no better reason than to support some very questionable policies of the United States. This begs the question of why the populace of the United Kingdom tolerated either of these conflicts. In a sense they did not. The massive demonstrations that ensued attracted – rather unusually – people from all walks of life and all political backgrounds but the government could, and did, simply ignore the demonstrators.

In a democracy, the people get opportunities to sack the government at election time, but those opportunities are relatively rare and an election is virtually never about a single issue. More to the point, how will changing the government help if the only two parties that can possibly win an election are both in favour of pursuing a war policy? This is particularly difficult in Britain, where the electoral system is far from democratic and massively favours two parties at the expense of all other voices. The dictum that a war cannot be long pursued if it is contrary to the wishes of the people sounds very reassuring, but it is not at all clear that it is true now – or indeed that it ever was. Perhaps, if the people of a democracy feel strongly enough about it for long enough, the war will eventually be ended by one or other political party – or perhaps a new party – which embraces peace as a policy objective. That may take years, and once a war has been initiated it is not always possible or practical to abandon it at a moment's notice, though generally that has been a rather threadbare excuse utilised by governments that have recognised the will of the people but have not wanted to carry it out … which is a pretty common condition in both war and peace.

None of this should be taken to imply or suggest that 'the people' are always implacably opposed to their country going to war; it is not always an unpopular policy and we should not ignore that fact, but it does call for examination. At different times, nations have gone to war amid a positive clamour of public excitement. In 1861, droves of men volunteered to fight for the Union or for the secessionists in America, and in 1914 British recruiting officers were overwhelmed by queues of young (and not so young) men eager to do their bit for King and Country. In the former instance, men volunteered for a variety of reasons – to save the Union, for States Rights or to end slavery; in the latter, men volunteered to protect 'plucky little Belgium' or because they saw service as a national duty. In both cases, the majority of men did not enlist or even give it serious consideration, but rafts

Vice-President Lyndon B. Johnson, Saigon, May 1962. Following Kennedy's assassination, Johnson used the spurious Gulf of Tonkin incident as an excuse to apply military force in Vietnam without consulting the Senate. Because he was a hawk? Rather, because his Republican rival for the presidency, Barry Goldwater, was seen as more hawkish by the public. 'All the way with LBJ', from 16,000 US military personnel in Vietnam at Kennedy's death to 550,000 by 1968.

of people staying away from the front does not make for dramatic newspaper material, and drawing attention to a reluctance to enlist does not suit the purposes of the government of the day.

Both of these conflicts – and many others – attracted men because they wanted to go to war. They did not volunteer to live uncomfortably in camps or to contract diseases to which they would not have been exposed at home and they certainly did not volunteer in the expectation of being maimed or killed or of suffering brutal treatment as prisoners of war. All of these were very real possibilities, but they were not what the men signed up for. However many chose to serve for ideological reasons – including a sense of duty – a very large number enlisted in search of adventure. Most of them, and most of the men who joined for more creditable reasons as well – were confident that the war would be short and that they would be victorious ... it would all be over by Christmas. It would be both crude and unjust to say that these men were seized by a pathological bloodlust, but it would be naive to assume that they did not expect to take part in battle or that they did not understand that battle would involve both killing and the risk of being killed. Tim O'Brien wrote in his memoir *If I Die in a Combat Zone* of a later war:

> It was no decision, no chain of ideas or reasons, that steered me into the war ... Men are killed, dead human beings are heavy and awkward to carry, things smell different in Vietnam, soldiers are afraid and often brave, drill sergeants are boors, some men think war is proper and just and others do not and most do not care. Is that the stuff for a morality lesson, even for a theme?

In both the American Civil War and the First World War there was a near-universal ignorance of the business of war. In 1861, the Mexican war was already the better part of 20 years in the past and although it had involved a degree of indirect conscription through the use of existing State militia units, for most

Americans Mexico was a long way away and the war had had a limited impact on society. In 1914, the British Army had extensive experience of war, but every conflict since the Crimean War more than half a century earlier had been essentially colonial in nature and had been conducted by regular and reserve army units … and they had all happened at a great distance. The potential for death – and most who gave that any thought at all doubtless envisaged a neat and immediately fatal bullet at a moment of grand courage – was greatly outweighed by the allure of adventure and excitement.

We might, then, consider the business of going to war to have a 'chicken and egg' element. Governments – whether by desire or otherwise – go to war and men (and some women too, though there is a tendency to skip over that for cultural reasons) are content or even eager to get involved. Equally, since men and women join the military as a career option, governments have the capacity to wage wars when they see fit.

There are people – mostly, but not exclusively, men – who are inclined to like a scrap; a phenomenon that can be observed in football (or soccer) where there are certain teams who attract a greater proportion of men who enjoy a bit of a fight than the supporters of most teams. Not infrequently, those 'supporters' particularly relish encounters with other teams that attract a similar following. This is not, perhaps, limited to football, though one has to admit that rugby, tennis or lacrosse riots are few and far between. We should, perhaps, simply accept that some men have a propensity for enjoying violence, but they are certainly too few in number to provide the raw material for a war, so making a war is the business, responsibility and not infrequently the fault of governments.

That does not mean that war can always be avoided, or that a smaller war in the short term can sometimes prevent a much larger one in the future. War can be, and often is, forced upon an unwilling participant. The playground adage beloved of teachers that 'it takes two to start a fight' is worse than just plain wrong; it is intellectually incoherent, an insult to the intelligence. However patient we may be, however much we would rather avoid confrontation, most of us object to being punched and kicked. Sooner or later, if we have no means of escaping violence, we tend to have recourse to it ourselves even when we know that we cannot be successful. This applies to countries just as much as to individuals. The right to act in our own defence is generally seen as a natural, even laudable, thing; though interestingly the same teacher who tells a child that it 'takes two' to make a fight may well tell the same child to 'stand up for yourself' without seeing any moral or intellectual conflict in the two statements.

At its simplest, we might conclude that war itself is not inevitable; it is only inevitable if one party is sufficiently keen to have one. That does not mean that the aggressor necessarily strikes first, only that they develop a situation where the other party is forced to react. More often than not it is the defender who squeezes the trigger and strikes the first blow.

2. The Study of War: Making a Salad or Baking a Cake?

Why do we generally think in terms of the 'Art of War' as opposed to regarding conflict as a science? It is certainly the case that technology seems to exert a greater influence over the practice of conflict than at any point in the past, but that is a long way from war being even remotely close to being a science. There are no universal rules. What works effectively at a given place and a given time with a given scale and nature of military force may be less effective or even utterly disastrous in very similar circumstances. The practice of war is immensely complex. The range of factors involved at even the most fundamental level of two small parties of soldiers fighting to acquire an objective or to deny it to the enemy is so vast that conflict is simply not amenable to establishing the degree of universality that we expect to find in physics or chemistry.

The absence of 'constants' in the sense of specific phenomena that are always valid and therefore predictable and which can be expressed as unarguable facts is what makes conflict so very difficult to understand. There may be some value to Alfred Burne's phrase 'inherent military probability', but accurately identifying what is genuinely 'probable' is extremely difficult unless we have a good understanding of the totality of the military experience in general and a well-developed, detailed knowledge of the forces involved and the environment in which a particular operation was mounted. All three of these things are quite rare, and in combination they are almost unheard of among professionals, let alone among the politicians, historians, journalists and instant experts (the kind that have read one book and/or seen one television documentary) who are most likely to use the phrase.

With the benefit of a little studying we can see that what 'works' in one set of circumstances can be a miserable failure in another, but understanding why that should be the case is a different matter. We may be able to identify some of the factors that have contributed to defeat or victory in the past, but we will seldom, if ever, be able to really understand every aspect of every part of a clash between opposing forces, let alone genuinely understand the significance of each part of the picture.

If we accept that war is not a science, do we necessarily see it as an art? There are certainly aspects that are not dissimilar. A good deal of what happens on the battlefield is the outcome of decisions made by commanders which are not made on the basis of equations and maxims learned in training or on deductions rooted in the study of past conflicts. Like an artist, a commander may learn a lot from the study of the great masters of days gone by, but his actions will be guided by the nature and size of his forces, his wider experience and his own ideas of what will 'work' for a particular project. His choice of actions will inevitably be influenced by cultural, social and ideological conditioning that may or may not have a relevance to the project and which may or may not be valid; he may well not even be conscious of those factors, but they are still going to have an impact on his general approach to dealing with the challenges of the battlefield.

In that sense, war can be seen as an art; every artist is the product of their background just like anyone else. Their work may be an almost linear development of the pieces produced by their predecessors or it may be a revolutionary departure or even an outright rejection of the artistic tradition in which they grew up, but everything they have seen or read will have had some influence on their own contribution.

The concept of war as an art has some validity, but the analogy breaks down when we consider outcomes as opposed to processes. The artist intends to produce work of quality, but what constitutes 'quality' or 'success' is very much in the eye of the beholder. To a great many observers, Picasso's *Guernica* is a forceful and haunting depiction of suffering in conflict; to others it is a childlike drawing of a horse, a bull and a disembodied hand holding an oil lamp. For the military leader, success seems to be rather more clear cut; he wins or he loses. In fact, that is an oversimplification in the sense that it may be quite impossible to achieve outright victory, but the commander may have demonstrated skill and insight by avoiding defeat at the hands of a stronger opponent.

The chief similarity between the artist and the general is that both must combine ingenuity, imagination, innovation, judgement and perseverance with knowledge and commitment to achieve positive results. It is not always the case that either will have had the benefit of a structured education specific to their work. History provides many examples of both painters and fighters who have been very successful despite a complete lack of formal training. Genius can trump study in both professions, but in either case, a knowledge of 'the classics' and of technique is likely to come in handy.

War is not a science and is not entirely an art either, though there is a degree of congruency. Can we, therefore, accept that there is a 'theory'? In the strict interpretation of a scientific vocabulary there really cannot be a 'Theory of War' in the sense of a 'Theory of Evolution'. The work of Darwin (and others who are rather less well known) does not give us a blueprint that explains every last detail of the development of every species, but it provides more than enough material to allow us to 'join the dots' and see a general picture of the living world.

War theory does not give us quite so clear an insight, so what is it about and what does it do? How does it help us in our understanding of conflict as observers or our approach to conflict as practitioners? Since there are no perfect routes to success we might argue that there is no reality to war theory at all, but we can see that the study of military practice tends to lead to a better understanding of the battlefield and of war generally. Equally, we can see that the failure to understand some general principles, and to apply them in a manner that is rational in terms of the command-decision environment, is very likely to lead to a failure to win at best and to complete disaster at worst.

Theory of war can never be more than a helping hand of guidance – it will not win battles of itself – but unless we come to grips with general principles we will struggle to gain any useful level of understanding (it will never be truly 'absolute') of the battlefield – of why this or that commander chose to make certain decisions or to pursue or abandon a particular course of action.

Clearly this is – or at least should be – a matter of some interest to military professionals and historians as well as to the substantial segment of the population that happens to be interested in warfare. There is a wider dimension as well. Since our society expends so much money and effort on defence, there is a real value in understanding, at least to a modest degree, the nature of conflict and the practical constraints on soldiers and government who have to make decisions in relation to preparedness for war.

·In an ideal world, defence planning would be a consequence of identifying and understanding potential threats and applying that knowledge in the light of the economic and political environment. In fact, this is a luxury that is seldom, if ever, available to either the politicians or the soldiers. Both have to work within an envelope which includes a near-infinite range of cultural and social assumptions, diplomatic commitments and traditional preferences which may be well past their sell-by date and be of limited relevance to the here and now. War theory will not provide us with all of the answers to current and future military challenges; it may not provide us with a complete solution to any of the problems, but it can provide us with a starting point. When we embark on a description of the principles or theory of war, we are faced with an immediate problem: where do we start?

Few people will come to study conflict without bringing an extensive collection of baggage with them. If we want to study war we must presumably be interested in it and we will, in all probability, have read some military history and watched conflicts unfold on our television screens. Most of us will have already formed opinions about the relevance and significance of political, technological, moral and cultural factors and have opinions about the brilliance or incompetence of certain generals, kings and presidents, and also perhaps the utility (or lack of it) of different approaches to war and combat. These may or may not be useful, fair or relevant. We may have arrived at our conclusions through observation and deduction or through prejudice and wishful thinking, but our opinions will not have been developed on the basis of a rational, planned course of study. This applies as much to the professional soldier with years of training and experience as it does to the casual student, the specialist academic or to the political leader who might actually have to make real decisions about the organisation and funding of the military establishment and, of course, the decision whether or not to go to war.

Quite simply, when it comes to war theory, there is no natural or inevitable place to start. The abilities of the lower strata of military activity may define ultimate success or failure, but it is not that level at which the more significant choices are made; on the other hand, the major decisions should be influenced by a detailed knowledge of the abilities of the lower strata. Making war is a decision made at the highest level of political life, but unless the politicians have a reasonable understanding of the workings of war (and they mostly do not) they are prone to serious errors of judgement in preparation and in the application of force. Clearly, neither a top-down nor bottom-up approach is going to provide us with more than a small part of the picture.

If we really must have a metaphor, perhaps we should look to cuisine rather than art or sport and consider whether war is akin to making a salad or baking a cake. Either process might

The revolution might not be televised but the Vietnam War certainly was. How much information should a government at war disseminate to its own people?

involve a great many ingredients, but precise quantities and the order in which they are added to the process makes relatively little difference to the salad whereas (based on extensive personal experience of producing the inedible) it makes *all* the difference when it comes to baking a cake.

All of the ingredients of war are relevant to the process, and therefore to the outcome, all of the time. No aspect can really be understood in isolation. The best infantryman in the world cannot function without his comrades or the economic and industrial capacity of his country or without good leadership and planning. The most brilliant politician or general will struggle to win a war if the troops are not competent or motivated. War Theory has to reflect this. We will toil to understand the course of an engagement if we do not understand the weapons of the day, but that knowledge will not help us much if we do not understand how the weapon was applied or why a commander chose to fight in that location and at that time. If we do not have a grasp of the process of fighting, we will toil to get any real insight into the nature of a campaign, a battle or even a skirmish, let alone a war.

There are very real limits to the extent of our understanding of any action; we cannot interview every person involved in the fight or every person whose contribution was significant. Even if we could, we cannot be sure that their accounts would be valid even if they were entirely honest – and we cannot be sure that their accounts would necessarily be honest at all. We cannot be sure that we have a complete understanding of every application of every weapon, every vehicle, every feature of the terrain, every aspect of the climate or the effect of exhaustion, hunger, shock and morale; all matters of some importance.

This does not invalidate the study of warfare, and some factors are not always quite as important as they might seem; the baker of the cake does not need to know where the sugar or the butter have come from to know that they are essential to the business of producing a cake, but he does have to have some understanding of the relationship between the two ingredients.

That, in essence, is what War Theory is all about. It is not so much the knowledge of specific arms, practices, principles and military structures as gaining an insight into the relationship between those and myriad other factors, whilst bearing in mind that the whole process is conducted while somebody is trying to kill somebody else.

3. Violence and Governments

Part of the package of belonging to an organised community is that we entrust the state with a monopoly on the use of force. We may not always approve of the application of force by the state, but most of us certainly do not want the monopoly to be broken up or to be shared with institutions that are not directly under the control of our government. There may be minor departures which we accept, such as security guards at public events or in supermarkets, but even then we accept them only as functionaries of commercial enterprise, and the security guard is never the equivalent of a police officer. Even in countries where security guards can carry arms, it is rare for them to have powers of arrest. We may not like the government and we mostly do not like politicians, but we do (mostly) trust them not to wage war on the citizens and, other than in cases of self-defence, we certainly do not trust the citizen to 'make war' on his neighbours. If anyone is going to have authority to sanction violence it is going to be the state, not the individual, and it will be the state that controls the armed forces. Any alternative arrangement is generally called a coup.

In the 1960s, there was a fairly widely held belief that war is a direct product of the existence of the military. A song written by Buffy Sainte Marie in 1964 (and recorded to rather greater commercial success by the British folk singer Donovan) claimed that it was the 'Universal Soldier' who was to blame for conflict. If there were no soldiers there would be no wars. Simple as that. A similar view was expressed in a popular poster which asked, 'What if they gave a war and nobody came?'

The line was not new; it had been borrowed from an anonymous combination of work by Bertolt Brecht and Carl Sandberg. The resulting poem actually offers a very different proposition:

> What if they gave a war and nobody came?
> Why then the war would come to you!
> He who stays home when the fight begins and lets another fight for his cause should take care.
> He who does not take part in the battle will share in the defeat.
> Even avoiding battle will not avoid battle, since not to fight for your own cause really means fighting on behalf of your enemy's cause.

It would be easy – and perfectly rational – to dismiss both the song and the poster as childishly simplistic, but the general sentiment is easily understood and has become strongly embedded in certain trends in both left-wing and right-wing thinking. A song of the 1970s told us that 'Violence is caused by governments, armies, police force' (Here & Now, 1977) and any number of conservative writers and politicians at different times has sought to reduce military expenditure as part of a general belief in reducing the cost of government and thereby lowering taxes, and also a view – shared by many on the left – that the reduction of armed force in one country will be followed by a commensurate reduction in neighbouring countries due to an outbreak of common sense.

The concepts expressed in the songs and the poster may be attractive, but they are not in any sense valid. It would, of course, be impossible to prosecute a war if there were no soldiers and there would not be much of a battle if nobody turned up to take part, but it is the society that makes the soldier, not the other way around. There is a double cost; the soldier must be armed and paid, which costs money, but there is an opportunity cost too; the soldier is effectively rendered unavailable for other tasks in the community which we might, quite reasonably, see as more profitable, so why do societies construct armies?

Societies and communities, except perhaps certain small, isolated groups of people who are not in contact with others, feel a collective need for self-protection against other groups. In fact, this extends beyond the bounds of interaction between human beings. Even the most dedicated pacifist would accept that a group of people are perfectly within their rights to take action to protect themselves against a predator animal which could enter their home and eat the occupants, though curiously there is something of a tendency to disregard that right in relation to attacks by human predators.

In the distant past, the protection of the community did not require a formalised body of individuals whose sole or chief function was war. It is a long-established tradition that those who benefit from membership of the community have an obligation to defend the community. Discharging that obligation might be performed in a wide variety of ways – supplying food, paying taxes, making equipment, providing shelter – that do not include physical combat, but the duty is generally understood as being necessary for the overall well-being and even survival of the community.

At the most basic level of the small hunter-gatherer tribe, the duty would certainly be universal in a physical sense; when the tribe was threatened by a predator or another tribe, everyone would be expected to take part in whatever measures might be required for preservation.

Over the centuries, the successful tribe of hunter-gatherers would become either more sedentary or – in some cases – retain their nomadic existence but develop a more sophisticated structure that could sustain the tribe. In either case, the duty to protect the tribe as a whole was still universal, but the roles of individuals became more specialised. A proportion of the tribe, generally if not invariably men, made the transition from hunter-gatherer to hunter-warrior and eventually, in one form or another, to just plain warrior. There might be a range of non-combatant duties attached to warrior status, but the primary role of the select warrior group was essentially to fight. Almost without exception, the warrior group in developing societies would come to occupy a privileged position in the community. This might arise from admiration for those who were willing to risk their lives or because they were the biggest and strongest men or through a general acceptance that those prepared to take the greatest risk should enjoy reward, but even in societies where being a warrior was not seen as a particularly honourable calling, some level of privilege would accrue to the fighting man. In due course, the role of the warrior tended to move away from 'those who defend the community' to 'those who control the community'.

This was the basis of what we call (not very usefully) the 'feudal' system and which we see (again, not very usefully) as a European phenomenon brought about by the Norman diaspora, which established itself as the ruling class across much of western Christendom in the Middle Ages. The warrior class still had the responsibility for the greater portion of military service and for the military leadership and organisation for war of the wider community in the face of threats, but also became the chief beneficiary of the economy and also the maker and enforcer of regulations.

The warrior ethic was not, of course, 'the only game in town'. There may be a good argument that part of the true origin of a formalised warrior class lies in a perceived need for the protection of a particular social ethos. Guarding the future of the community might be seen as a rather more complex matter than simply retaining control of land for hunting and agriculture, but of 'preserving our way of life'. All societies seem to have developed some form of belief in the supernatural; an assumption that there is some sort of powerful superior being which controls or influences the destiny of the community and which must be 'protected'. The logic is weak – if the superior being is as powerful as all that, why would it need the protection/adulation of mere humans – but it does seem to have been nigh universal throughout recorded history and archaeological evidence would indicate that the belief in the supernatural existed long before any recognisable means of record-making. We cannot realistically accept the idea that the massive investment of labour required to erect monuments such as Stonehenge was made simply to tell the community that winter had arrived ... the snow on the ground might have given them a hint. We might therefore question which came first. Was it a belief in gods whose rules should be obeyed and who needed to be protected against the beliefs of a neighbouring community or was it the rise of a clerical class whose role and status as the 'hot-line' to the gods would be best protected by becoming an adjunct of and support to a warrior class with a powerful position in the community?

Regardless of the validity of religious belief or of the desirability of preserving a particular religious ethos, the step from 'protecting our way of life' as a community and 'protecting our way of life' as a means of furthering the fortunes of a specific class within that community is a very short one. When people talk of the power of the 'military–industrial complex' they are not altogether wrong, though it is important to remember that the rise of what we might term the 'ideological–military' complex is really the more significant development. If there was no economic opportunity in tending to the military requirements of the ideological portion of the equation there would be no incentive for industry and commerce to engage in the provision of arms or to pay the necessary taxes.

It is also a very short step from 'protecting our way of life' to imposing that way of life on others. Often, the desire to impose standards and practices on a neighbour has been based largely or even entirely on economic motives, and sometimes on ideological or theistic ones, though the economic aspect is seldom out of the equation.

Conquest which is attempted in the name of a god or a political persuasion has mostly been very strongly linked to the prospect of improving the lot of the aggressor. The aggressor may well bring a religious or political ideology to the vanquished, but the victor generally envisages some form of economic gain either for the community or for the individuals that lead the fight. The Crusaders may have genuinely believed that they were doing 'the right thing' by God when they conquered the land in which Jesus was born, lived and died, but it did not stop them from erecting new kingdoms on the European model in which – curiously enough – many of the crusading lords selflessly took on the burden of lucrative lordships. We might make the same observation about most, if not all, other attempts at conquest; the spread of Islam by military force may have brought the word of God to the masses of early medieval Spain, but it also brought immense wealth and privilege to the men who settled there.

Modern conflicts are not really very different in principle. Men (and it is still largely men, though less so now than in the past) who serve in war expect to be rewarded. This is not unreasonable in itself. Any of us who make an investment of any kind – finance to start a business or time to learn a skill or risking one's body on the battlefield – do so in the hope and expectation of benefit. The reward may not be economic, but whether it is the satisfaction of producing an artwork, or persuading others of the value of a certain philosophical position, the reward is what motivates us. In war, the personal reward varies considerably. Men who go to war no longer do so in the hope of acquiring a landed estate or a string of slaves. Their reward may be as simple as the wage they receive or the expectation that they will gain rank and status through the quality of their service in the field. They may be motivated by social, political, racial or theological beliefs, in which case their reward may have no physical manifestation at all, but that does not mean they fought in a spirit of altruism.

Various societies throughout the ages have believed that their practices were superior to those of their neighbours and that there was something to be said for extending the benefit of those practices at the point of the sword or the muzzle of the gun. It would be a curious thing if their neighbours did not feel that they too had a lifestyle and system worth preserving and therefore took steps to defend themselves. The mere fact that they have a means of protection – an army – is not evidence of a desire to go to war, only of a determination not to be at the mercy of a neighbouring country or society. The existence of the army, then, is not the cause of warfare so much as a means of preventing war from being the engine of defeat and subjugation.

We might take this a step further and ask whether war is actually the product of social development rather than an impediment to it. The 1970 hit 'War. What is it good for?' is a laudable indictment of conflict, but the reality is somewhat different. War is not always the only means of combating the aggression of others, but it is sometimes the only route available. It is certainly true that there would never be a war if absolutely nobody wanted one, but the sad reality is that there is generally someone, somewhere, who does want one and enough people who are at least willing, if not eager, to help. Everybody who ever advocates a war can provide themselves with a good excuse – or at least an excuse that satisfies themselves if nobody else. Having done so, the onus is then on their targets to rise to the occasion and defend themselves. The assertion that it is 'better to die on your feet than to live on your knees' is somewhat trite, but it is not completely unreasonable.

We might ask whether a minor change in our society imposed from outside is really enough of an issue to make a war 'worthwhile', but the change may not be minor; the enemy's objective may be genocide. We might, in any case, ask what constitutes a 'minor' change. An attempt to impose a different religion or even a marginally different variant of the same religion may seem to us to be a trivial matter, but to the people on the receiving end it may be a matter of whether they not only die in the here-and-now but also suffer unspeakable torment for eternity.

The decision to defend a particular form of government or social practice may be very much greater than a question of whose backside rests on the seat of power. It may – and often will – have far more significant consequences and may call for a greater commitment than what we might strictly call 'defence'. By the close of 1940, it was clear that Hitler's ambition to conquer Britain was no longer an immediate threat to the continued existence of the United Kingdom either as a nation or even as the hub of a worldwide empire. Some political figures on both the left and the right believed that reaching an accommodation with Hitler would be the best course of action since the country had clearly managed to defend itself. In a very simplistic sense they were, of course, correct. The immediate threat of conquest had been averted and the expense of building and maintaining large military forces was an immense burden on the state. Some of these political figures were, essentially, driven by the concept of 'if we do not bother them, they will not bother us' and others by a view that fighting is always wrong or by a simplistic, literal view of the sixth commandment, though 'thou shalt do no murder' is a more realistic translation than 'thou shalt not kill'. The wider popular view was rather more realistic; 'if we do not work to stop the Nazis overseas, we will have to fight them in our own streets and fields.'

This was a practical analysis at that time and in those circumstances, but that sentiment has been a tool for governments which – for whatever reason – want to conduct a war with a less self-evident rationale or even with no real, practical rationale at all. Both of the main parties in the United Kingdom have used it to defend the recent conflict in Afghanistan as a 'war on terrorism', as though there was every chance that failure to defeat the Taliban would lead to bombs and firefights on British streets. In practice, the Afghan operation has, if anything, made Britain more of a target than would otherwise be the case, and has done so at the cost of hundreds of British servicemen and servicewomen killed, thousands wounded and billions upon billions of pounds dispensed from the public purse. In the absence of any clear objectives gained, it is hard to argue the sacrifice has been worthwhile.

Given the undoubted need for countries of like mind to stick together, there might have been a good argument for British involvement; however, the real objective was for the British political leader of the day (as opposed to the government of the day) to lend support to his most important political ally, more a matter of 'Tony loves George' than 'UK loves US'. It is not even clear that there were any real strategic benefits to be gained from American involvement either; some uncharitable souls have suggested the root cause was the inability of George Bush Junior to distinguish accurately and consistently the difference between Iran, Iraq and Afghanistan.

This is not a new phenomenon; it is not unknown for the leader of a country to conflate what they see as a good thing for their own political career, beliefs or historical legacy with what is good for the country they lead. There may be a real relationship, of course; Churchill's historical legacy will always rest on his leadership in the Second World War rather than his advocacy of state-funded old-age pensions and social insurance for the benefit of the working class. Unlike Blair, Churchill's objectives were perfectly clear and straightforward; the defence of his country and the preservation of her status as a great world power. One might question specific policies and the success of his attempt to ensure that Britain remained the powerful force that she seemed

President Obama has a heart-to-heart with General Stanley McChrystal,
Commander, US Forces Afghanistan. Was McChrystal recalled to Washington in June
2010 because of incompetence? No, it was because he criticised some of the civil
administration in a magazine article and thereby, indirectly, the Chief. War is hell but
politics is merciless.

to have been in the 1930s, but his primary motivation was the
good of the country.

War may be a product of political development or political
development may be a product of war – whether defensive or
offensive – but it has an incredibly wide range of effects on virtually every part of our lives. War was not the spur for the invention
of the aircraft – Wilbur and Orville were motivated by a mixture of economic ambition and scientific curiosity – but it was
unquestionably the greatest factor in the rapid development of
powered flight. In 1903, the Wright brothers made history with
a powered flight of a few seconds over a few yards. Sixteen years
later Alcock and Brown made a flight across the Atlantic Ocean;
they did so in a converted military bomber. Advances in engineering, chemistry, physics, medicine and many other fields would,
undoubtedly, have been made in the absence of war, but at a frac-
tion of the speed.

Of all the progress made in time of war, we might consider
the fiscal ones to be the most important. There is an argument

that early improvements in agriculture have their o
demand for rents in produce (and later cash) from peasa
ers to support warrior-class landlords. The rent of property b
fixed as a certain volume of produce led to efforts by the farme
to get more out of the land and thereby secure a bigger surplus,
which could then be traded. As the requirements of the warrior
class became more sophisticated – armour, better armour, a horse,
a better horse, a horse with armour and so on – the demands on
the farmer became heavier, so he brought more land into cultiva-
tion, improved the livestock by selective breeding and learned the
value of manure so that he could pay his rent and still maintain
the same standard of living, or even improve it. Few of us tend
farms, and virtually none of us is actively involved in agricultural
development, but the demand on our income to support the military establishment is still with us. When William Pitt introduced
an income tax in his 1798 budget – nearly 1 per cent from the
middle classes and an eye-watering 10 per cent from the very
wealthy – he did so to pay for war. Today, we expect to pay a
good deal more in the way of direct personal taxation and far
greater sums in indirect taxation (which, of course, still comes
out of our pockets eventually) and we expect the money to be
applied to a much wider range of activities, but paying for war, or
at least the ability to wage a war in the future, is still a major drain
on our resources.

Complain as we might (and most of us do) about the endemic
incompetence and dishonesty that is part and parcel of entrusting
our government to political parties led by people whose only
real skill is that they have been able to get to the top of political
parties, the fact remains that we all expect our government to
ensure our physical safety. This is true whether it is a matter of
protection from criminals at home (though not, as a rule, from
the ones at the top of political parties) or hostile elements abroad.
It is, in fact, the primary duty of government: the preservation
of the country. Like it or not, governments really exist to protect

itself, may sometimes require the physical
[...] e are under threat it may be that the best
[...] r safety is to take action before the threat
[...] resist.

[...] the best defence is a strong offence has its
[...] most clichés, it also has more than enough
[...] n validity. It is quite possible that adequate
and effective defence may require offensive operations far from home, but how do we identify the difference between defence at a distance and projection of force for ideological or economic purposes? Does our nation have a Ministry of Defence or a Department of War? Is there really a distinction? For practical purposes we should, perhaps, assume that there is none at all. The purpose of the military is either to deter threats or to destroy them. If the threat cannot be deterred and the situation is not susceptible to reasonable negotiation, it must be dealt with by direct action, and at that point we inevitably have to consider the possibility of fighting. Most governments are aware that, in the end, in the words of Thomas Hobbes, 'The obligation of subjects to the sovereign is understood to last as long, and no longer, than the power lasteth by which he is able to protect them'.

4. Politics by Another Means

War is a means of attaining objectives that cannot be acquired by political negotiation, or a means of preventing an opponent from doing so. Wars have several territorial dimensions. The traditional view that war is essentially a path to conquering territory to increase the resources of the aggressor is not the only one.

As we have already said, a war may arise from any number of other factors, such as a desire to impose a particular political or

religious or racial policy on a neighbour or to defend against
such an imposition, but such courses are very often defined by
the attempt to gain control of the landscape. Our general understanding of what we might call 'formal' warfare – the process of
two opposing armies manoeuvring and offering or forcing battle
to achieve territorial objectives – rather supports that proposition, but it is not always the case. Control may be achieved
through success on the battlefield, but it can also be achieved
through economic attrition. The party with the greater wealth
or motivation, or at least the ability and will to mobilise those
factors to a given purpose, may be able to drive its opponent
to accept conquest without winning encounters in the field.
Years of conflict in Vietnam did not result in major victories for
the North or major reverses for the South until the departure
of the Americans; victory was gained through – among other
things – the greater willingness of the North to pursue the objective. Contrary to general perception, the Vietnam War was not
conducted as a guerrilla conflict. There was certainly a guerrilla
element, but the efforts of the Viet Minh (the term 'Vietcong'
is an American propaganda invention) were always subsidiary to
those of the army of North Vietnam and were of limited value.
By and large, it was a formal conflict in which two very large
forces fought and manoeuvred in a fairly conventional manner
rather than an insurgency.

The traditional view of an insurgency movement is an assumption that the insurgents will start with minor operations against
weak or isolated elements of the opposition and – as the movement grows in strength – take on ever-larger operations until
such time as they can secure territory and set up an administration. As time goes on, they will enlarge that area until such
time as they become the effective government of the whole state
or, in the case of nationalist movements within an existing state,
that portion which they consider to be their natural borders –
whether on racial, linguistic or religious grounds – or perhaps

restoring the political integrity of a nation which has previously been lost through conquest or political manoeuvres.

This is, in essence, the Maoist model and a great deal has been written about it, but it does not have universal application. Not all insurgencies aspire to creating a new state. Some are specifically geared to transferring the sovereignty of a particular region from one state – who the insurgents see as an occupying power – to another, which, they believe, for whatever reason, to be the rightful 'owner' of the region in question. The Maoist model could be applied in those circumstances but the insurgents may take a rather different approach, that of making the region so difficult to police and such a drain on the wealth of the 'occupiers' that at a given point all parties come to the conclusion that they will be better served by negotiation than confrontation. This course of action is only a viable outcome when all of the parties come to accept that they cannot bring about victory through conflict and when some – or very often all – of the parties are prepared to make political concessions. The extent of the concessions made will generally reflect a variety of factors, which may or may not include military advantage. If there is no clear sign of progress, the operatives and supporters – of one or all parties – will eventually lose confidence in the project. The parties may lose whatever sympathy they have enjoyed in the local community or in the eyes of the rest of the world, or the people who have to foot the bill may lose the will to keep paying the price.

It is not always, or perhaps even generally, possible to define a conflict as being completely 'formal' or 'insurgent'. Even the most conventional of campaigns is very likely to include elements that embrace the kind of practices that we associate with insurgencies. Small parties of specialist troops operating behind enemy lines, which engage in sabotage and assassination, are no different in principle from guerrilla bands. The chief difference between partisans, terrorists, freedom-fighters and bandits is the manner in which we perceive them. If we favour one cause over another, the

soldiers who are dropped by parachute or from boats in the dead
of night to blow up rail junctions or bridges are heroic figures
carrying out dangerous duties for their country. If we favour the
other side, they are merciless terrorists. In either case, the activities have much the same range of combat values and little, if any,
difference in application.

Such operations can only be frustrated by deploying considerable resources in troops and money which could otherwise
be devoted to the main arena of the conflict. They have much
the same impact on the wider society. For a time, the 'folks back
home' may actually become more motivated by the intrusion of
the raiders, but they may also become less confident in the ability of their government to protect them. There is nothing new
about this aspect of warfare; it is not a product of modern ideological development or military technique, but it is all too easy
to assume that that is the case. In fact, our general understanding
of the history of various conflicts can – and often does – lead to
very questionable conclusions.

Taking the Hundred Years' War as an example, we might – and
mostly do – see it as a conflict dominated by a surprisingly small
number of dramatic engagements: Crecy, Poitiers and Agincourt.
More widely, we might identify a process of besieging and capturing significant towns and castles … 'once more unto the breach'
is a phrase with which we are all familiar. In fact, the day-to-day
business of the war was really about imposing one administration at the cost of another – or even just preventing an opposing
party from exercising government. To a very great extent, this was
achieved through the operations of rather small parties of mobile
troops demanding rents, taxes and produce from communities
and thereby imposing the rule of French or English kingship.

In principle, it was not different from the Maoist model,
though the vast majority of the combatants were not the peasants
or the factory workers of Mao and Marx fulfilling their destiny
in search of equality and liberty, but the gentry and aristocracy

English archers from the Beauchamp Pageant. The Hundred Years' War is remembered for set-piece battles such as Agincourt but most of the conflict over such a long time span comprised harrying raids and coercion of the local populace.

fulfilling their obligations to their king and looking for wealth and glory in the process. The acquisition of castles and towns brought the ability to impose government, but the significant military activity was, in essence, raiding.

Raiding, therefore, has several objectives. Tactically, it encourages the enemy to use resources that could be better employed elsewhere; economically, it requires the repair of industrial or other assets which the raiders damage; and politically, it may undermine the morale of both the wider community and the troops at the front, who naturally will have concerns about their families and friends.

This aspect of conflict is not, of course, limited to the clandestine operations of soldiers in small groups operating in enemy territory. The suicide bomber who blows up a supermarket and the disaffected civil servant who quietly undermines the workings of their department are both really doing exactly the same thing as a comrade hiding out in the forests or mountains awaiting an opportunity to shoot an enemy soldier.

The same can be said of various aspects of what we generally perceive as part of the more formal approach to war. Aerial bombing of factories and power plants or commerce raiding at sea are not different in principle. The objectives are identical; disruption of trade and industry, undermining morale at home and at the front and, of course, forcing the enemy to divert troops and material to purposes which are not, superficially, relevant to the main thrust of the war.

The key word is 'superficially'. In reality, every part of the conflict is a crucial part of the business. The great military writers – including Sun Tzu, Christine de Pisan, and Carl von Clausewitz – all understood that a key element in attaining success is making the enemy react to your actions. It is not even necessarily vital that the actions undertaken should be particularly favourable to you, only that they are unfavourable to him. If the entire business of war boils down to resources and commitment, everything and anything that diverts the enemy from pursuing his objectives is desirable and should be exploited so long as it does not demand a level of effort which undermines your own endeavours or makes unbearable demands on your own support. (Consider the assertion of higher command that, when sending the Chindits into the jungle in the Second World War, it was a *positive* that disease was inevitable – that they were entering an environment that was inimical to life itself. The Japanese would have more troops there to counter the incursion than there would be Chindits, so would suffer more, like for like; a horrible calculation.)

All conflicts are supported by the efforts of the communities involved and the willingness of that community to bear the cost. How willing they are depends on a great many factors. The community may be strongly motivated by political or religious ideology or by the prospect of economic gain or by survival, but no war can be waged endlessly without some degree of willingness to suffer the consequences. In the short-to-medium term a community can be persuaded or coerced, but unless victory can be attained there will eventually be a weakening, and eventually a failure, of the will to fight on.

Any conflict has an economic cost for all parties. National governments may be willing to commit everything to the struggle. They might sacrifice thousands or even millions of lives, run up enormous debt and incur massive damage to the social and economic structures of the nation. They might feel that the costs are outweighed by the potential benefits or they might be obliged to pay the cost to prevent their country being conquered. They might want to be involved in a conflict to ensure that they have influence in the political and diplomatic settlement when peace is restored. There is a powerful argument that this was a crucial factor in Britain's entry to the First World War in 1914. Whatever the cause and regardless of whether the result is defeat or victory, a price has to be paid. In a democracy, the government that takes the country to war can always hope that the price will be the political responsibility of another party, but the money has to come from somewhere and the populace will have to find it in higher taxation, higher prices, reduction of services and amenities or, as is often the case, all three.

The same holds true for less formal belligerents. Insurgents may be able to draw on the resources of another party with similar political, religious or cultural sympathies, or which has an interest in undermining the power and stability of the counter-insurgents, but sooner or later they will inevitably have to make demands on the community which they represent – or at least claim to represent. If the insurgency is genuinely popular, they may receive freely given material support from the community. If not, they might take what they need at gunpoint, but there is inevitably a cost to individuals or groups. There will also be a more general burden. If the insurgents blow up bridges and factories, the community will – as a rule – have to pay for rebuilding or suffer unemployment. If the insurgents choose to kill police officers as 'instruments of the occupying power' there will almost certainly be an increase in criminality as the resources of the police force become more strained.

Equally, the counter-insurgency may decide that the effort involved in maintaining the status quo is greater than any economic or political advantage they hope to maintain or gain, and be driven to making the concessions that they were fighting to avoid – which is, of course, the outcome that the insurgents hoped to attain in the first place.

There may be other consequences that were not on the agenda of the belligerents. The First World War did not instigate an extension of the franchise in the United Kingdom, but it certainly accelerated the process, most particularly in regard to votes for women. The nature of governments often – perhaps always – tends toward introducing reform gradually as though there was something inherently desirable about preventing change. Once the franchise was extended to some women, it was inevitable that the same voting qualifications would apply equally regardless of gender; even so, it took more than a decade before the British government made the necessary reforms.

Unexpected and unintentional developments are nothing new. When Robert Bruce set about making himself king of Scotland he had no intention of waging a war that would eventually bring about the end of serfdom – indeed he would enact legislation requiring runaway serfs to return to their owners – but that was part of the wider outcome of the wars of independence. Servile status was certainly in decline across Europe and the upheaval and dislocation of a protracted war had an impact on the opportunities for individuals to change their status, as was the case with conflicts in other countries. The political dimension, however, was that it was difficult to retain the support of serfs in paying for a conflict for national liberty if they were not going to be personally free. As a result, servile status had disappeared from Scotland by the 1360s but would linger on for another 200 years and more across the rest of Europe.

We might make similar observations about other conflicts. Chairman Mao did not envisage China as a capitalist powerhouse that would dominate manufacturing for the whole world, and Neville Chamberlain did not commit Britain to war in 1939 assuming that the conflict would bring about the demise of the British Empire; what the belligerents seek is not necessarily what they achieve, even if they are successful on the battlefield.

Consequences are not always as unexpected as belligerents might suggest at the outset of hostilities. Abraham Lincoln identified war as the only means of preserving the Union and claimed that if he could do that by freeing all of the slaves, or none of them, or by freeing them in some places but not in others, he would take the necessary steps. In reality, though many Union soldiers had no opinion on slavery and many were actually opposed to abolition, the war was most certainly about slavery, and a Union victory would unquestionably bring the 'peculiar institution' of the southern states to an end. In 1967, Israel did not fight a war specifically to gain territory, but the acquisition of the Sinai and therefore of a buffer area between Israel and Egypt was rather more than just a tactical and strategic accident.

The policy considerations that bring about conflict may emanate from ideology, threat or expansionist ambitions and are the province of the political class, but someone has to plan and execute the operations and someone has to do the actual fighting, not to mention the dying. There must be systems and structures to enable the combatants to carry out the wishes of the government, none of which can really be understood, even very superficially, unless we look at practical applications in tandem with theoretical concepts; and none make any real sense unless we can arrive at some understanding of the pressures, challenges and workload of those who find themselves face to face with the enemy.

PART 2

THE PURSUIT OF CONFLICT

5. Snow White and the Difficult Rifle Section

Regardless of why or where the fight occurs, somebody has to make the strategic and tactical decisions, somebody has to set the operational policies and parameters and somebody has to do the actual fighting and dying. Senior officers do their best to preserve their forces for obvious reasons, but sooner or later pursuing a plan is generally going to lead to men engaging in combat. Every battle is, in the end, a conglomeration of minute actions in which at any

The First Rifle Practice of the Women's Reserve Corps

Messenger: The Major wants to know will you please excuse him from reviewing the practice to-day. He doesn't feel strong enough

A monstrous regiment, 1917. The Women's Reserve Corps' rifle practice frightens off the Major. In January 2013, the US military formally ended its blanket ban on women serving in frontline combat roles. The American Civil Liberties Union, which filed a suit in November 2012 seeking to force the Pentagon to end the ban, was delighted that the right to die on an equal footing had been won.

one moment one person is endeavouring to kill another person without getting themselves killed or wounded in the process.

Clearly, a huge number of armed soldiers wandering around the countryside taking potshots at other soldiers cannot be an effective approach to waging war; there needs to be an organised and disciplined approach to battle. More than half a century ago the American writer, soldier and analyst General S.L.A. Marshall wrote that all battles hinge on the successful actions of the smallest sub-units of armies.

This is not, perhaps, quite as obvious to the casual observer as it should be. It is all too easy to become fixated on the merits of a fast aircraft or powerful tank and lose sight of the fact that the competence (or otherwise) of 'boots on the ground' is actually the final arbiter of success in battle. If a location is to be considered genuinely secure, it must be a place where the enemy cannot operate. Short of poisoning the land with radioactive or chemical weapons, the tool for achieving security is the infantry soldier. Every aspect of the military structure boils down to the business of putting an infantryman into position, providing him with the means of conducting combat and giving him every possible advantage in the way of armour, artillery and aviation support that can be mustered so that he can meet and defeat the infantry of the other side.

Most professional soldiers are not especially eager to go into combat; peacetime soldiering is undoubtedly somewhat dull, but it does have certain advantages – longevity being one of them. This does not mean that there are not individuals who just like to fight – there most certainly are – but they are few and far between and the most extreme among them generally need to be kept in secure confinement to protect the rest of the community. The military life is not, as a rule, attractive to such people because they will not or cannot conform to the discipline and self-restraint required of the soldier.

By and large, the military consists of exactly the same sort of people as one would encounter in the wider community, but

Boots on the ground. Soviet infantry pulling a Maxim machine gun forward during the Battle of Kursk. It is the poor bloody infantry that wins battles. (From Mark Healey's *Zitadelle: The German Offensive Against the Kursk Salient 4–17 July 1943*, The History Press)

brought together, welded into a team and trained for a particular purpose. They cannot act as individuals if they are to be successful in battle, so there needs to be a team structure and somebody needs to be the team leader.

The weaknesses and strengths of General Marshall's research have been examined by a multitude of authors – notably in John Whiteclay Chambers's paper for the American Strategic Studies Institute – but the basic premise is sound: 'platoons seal the fate of armies.'

In turn, the fate of the platoon depends on the competence of the sections (or squads) of which it consists, and the abilities of the men who lead those sections determine their effectiveness in battle. No level of command requires interaction with more individuals than the leader of the basic building block of every infantry unit so we might do worse than have a brief look at the challenge that section – or squad – command presents.

More often than not, the section leader is a man of fairly ordinary intellect who has hopefully had the benefit of extensive and highly focused training. In theory at least, that training is based on an accumulation of experience gathered in both peacetime and wartime. It has generally been the focus of detailed examination and analysis and has been delivered by skilled officers and senior NCOs to give the section leader every possible assistance in his own professional development. If the section leader is a committed and enthusiastic person he will have made his own observations and will also have learned how to apply the lessons learned in the past to his own particular section members so that the whole group functions as well as it possibly can in the heat of battle; and also in the rather more tedious business of soldiering away from the battlefield.

In peacetime, the squad leader has to keep the private soldiers busy – if only because if they are not busy there is a much greater likelihood that they will find some form of trouble to get into. Just as importantly, the section must be kept in tune with one another and in a good state of fitness and training. This is more difficult than it might sound. Training must be repetitive in the sense that the drills and practices of battle are to be kept up to date and honed to a high standard, but not so repetitive that the soldiers become terminally bored and lose focus. If peacetime soldiering is difficult, service in combat is – not surprisingly – even more so.

Our chosen section leader is Snow White. Previous experience as a lower management figure in a non-specific mining operation has not prepared her for combat leadership as such, but it has allowed her to demonstrate abilities that will stand her in good stead. Any criticism of her absence from the 'pick and shovel' aspect of the work has been offset by her clear commitment to ensuring the well-being of her workforce – clean and comfortable billets, good food and the ability to motivate the entire team to accept her leadership. Her popularity has been demonstrated by the reaction of the group to her near-death experience in a fruit-related incident. Her failure to spot a rather obvious saboteur is bound to have been a lesson learned for the future. Admittedly, her capacity for musical communion with small woodland creatures will probably not be a great asset on the battlefield – though *any* knowledge of the theatre is not to be dismissed out of hand. Since making a career transition to the army, she has done well and has passed through at least one junior leaders' selection and development course with good grades. Her first command consists of a group of soldiers whose names – by good fortune – happen to reflect aspects of their personality.

Cheery, Nasty, Psycho, Lazy, Noisy, Clumsy, Bossy, Dimwit and Stinky have all passed through their initial training and infantry school without serious mishap and are all, as the army sees it, trained soldiers ready for combat. In the sense that all of them are reasonably familiar with the weapons of the section – which for our purposes might mean assault rifles, a light machine gun and a variety of grenades – and all of them have taken part in a series of training sessions relating to their roles in combat, and in several drills and exercises. The value of their training is obviously somewhat dependent on their own abilities and physical attributes; Noisy – to take an example – is always going to struggle with a fast run due the fact that he has very short legs and is, to put it politely, less than sylph-like. Snow White has eyes in her head and is not stupid, so she can take account of Noisy's inability to run fast. Whether she can ever get him to overcome his uncanny ability to make an outlandish amount of noise in any and every circumstance is a different matter, but we can be confident that she will do her best. In the early stage of the relationship between Snow White and the section, she will have the opportunity to analyse and understand the nature of the people who gave the section their basic training. She might find that the standard is very high or she might find that the eagerness of training camp NCOs to get the troops through the process has led to the cutting of more corners than she can count on her fingers and toes. She will have to deal with that in addition to her other duties.

If she is fortunate, the section will consist of reasonably well-motivated individuals. This is more likely in a peacetime non-conscript army. If the recruits were not attracted to the military life they would not have volunteered in the first place. If they were ill-suited to the life they would hopefully have been weeded out in basic training or in that first battle school period ... or so one might think. An obvious issue with that is the individual who really is not suited to the army, but likes it. Unless he is really, really hopeless, there will be a tendency for training staffs to make allowances (or ignore inadequacy) so that a well-motivated soldier is not lost to the service. This is not completely unreasonable. Some people take a while to find their feet, but make good soldiers in due course. Alternatively, the not-very-useful individual might pass through the training process for less desirable reasons. It is a challenge to keep infantry units up to strength. The proportion of recruits who find their way to the infantry is not that large and the army may simply have to take what it can get. There is not too much that Snow White or the training staff can do about that. What is worse is when men are carried through the system because the training staff does not want to be seen as failing to provide the quantity of men required, or are unwilling to admit that they found a certain individual to be untrainable (a little like school 'exam inflation').

As it happens, Snow White is in luck. All the guys are competent with their weapons, can carry out the drills and are reasonably fit. Now she has to make them into a team that will function competently in battle.

The temptation is to concentrate her efforts on Lazy, Clumsy and Dimwit and to rely on the willingness and popularity of Cheery. This may not be wise; if nothing else, it could easily lead to the demoralisation of Cheery if he feels that he is consistently having to do a bit more than everyone else.

However hard she tries, she is never going to raise the IQ of Dimwit; it may not even be desirable to try. As he stands, Dimwit is keen and has no trouble obeying orders just so long as he can remember them long enough to carry them out. That may not be such a great problem since he will generally be in the company of at least one more soldier who can keep him up to speed. To some degree, the same applies to Lazy. One way or another, he will manage to get away with doing a bit less work than anyone else, but once Snow White is aware of his nature she can take some steps to ensure that he does rather more than he would really like; his comrades can be counted on to do the same, so with any luck his idleness will not become a source of continual resentment for the rest of the section. No enemy soldier is ever going to fall victim to Clumsy's stealth, but he has other virtues such as courage and reliability which – with any luck – Snow White will be able to develop. He has the potential to be a real asset to the section though perhaps he should be discouraged from handling grenades.

Although each of these soldiers is a challenge in their own right, the real problems are likely to come from Nasty, Bossy and Psycho.

Psycho has some fine soldierly qualities; primarily that he is fearless and always ready for a scrap. The downside is that he is fearless and always ready for a scrap. The latter may lead to punch-ups in the barracks, which is clearly undesirable, and especially so if he is very tough. Other soldiers will come to fear and resent him, which is not conducive to a good team. His eagerness to engage in combat may well make him unpredictable and uncontrollable in action – a tendency that can easily put the whole section at risk.

Nasty has performed perfectly adequately in training and can be relied upon to carry out instructions to a reasonable standard – maybe even a good standard – but he is a bully by nature and will find ways to make other soldiers miserable simply because it entertains him to do so. Poor old Stinky – a good sort but you would not want to stand too close to him – is particularly vulnerable. Some of the section will have sympathy for him when Nasty is giving him a hard time and others will feel that his predicament is his own fault and could be solved by a better acquaintance with soap. Most of them – perhaps all of them – will just be happy that Stinky is the target and not themselves. Either way, Nasty is grit in the system. His ability to generate an unhappy atmosphere will very probably outweigh his value to the section as a combat unit.

Bossy is, however, the most pressing problem. If he gets away with throwing his weight about behind Snow White's back he will undermine the confidence of everyone in the section, with the possible exceptions of Psycho and Nasty. If he is unsuccessful at throwing his weight about he will inevitably become a target for Psycho and Nasty, but is most unlikely to get any sympathy from anyone else. Over time, he may become bitter and disenchanted and be a liability rather than an asset. A big problem with Bossy is that he may undermine the authority and credibility of Snow White. This is that much more of a problem if the other members of the section believe him to be more intelligent, more resourceful and a better leader than she is. It becomes an even bigger problem if it turns out that he actually is (or even just appears to be) smarter and more savvy. Bossy is very likely to get opportunities to enhance his standing by pointing out real, questionable and downright imaginary shortcomings in Snow White's abilities; all the more so since he will not, as a rule, be called upon to show that he could have done any better. Increasingly, the section will look to Bossy to make decisions regardless of what Snow White has to say.

Just because he is Bossy by nature does not mean that he cannot be a useful member of the team; in fact, quite the reverse. If Snow White can get Bossy on side, she may be able to do better than make him a good soldier; she may be able to turn him into a useful subordinate who can be trusted to keep an eye on things if she is absent for any reason, or to take a leadership role as and when the section operates as two teams. There are different approaches that Snow White could use to get Bossy to be a help rather than a hindrance. She can threaten him with the intervention of higher authority, though that should be the weapon of last resort. If she is seen as running off to the platoon commander with every problem, she is going to struggle to assert her own leadership. The preferable alternative is to recruit Bossy rather than marginalise or overpower him. Bossy is probably reasonably bright and an appeal to his intelligence may be the place to start: 'I realise you have ideas, but somebody has to be in charge and it's me; the team cannot work well if someone is undermining me ...' Another course is to appeal to his self-interest: 'I realise you have ideas and that you are not stupid. One day you, too, may well be a section leader, but only if someone recommends you for promotion. That someone has to be me ...'

Of course, it is always possible that Bossy will not respond to any form of persuasion or the threat of higher authority and will continue to be an unmitigated pain. It might prove possible to beat him into submission by finding endless tasks to fill his time and giving him as many dirty jobs as possible whilst still holding out the prospect of co-operation and an easier life. An undesirable approach – but one that can bring very positive results – is to make sure that everyone learns to loathe and detest Bossy. This can be achieved by finding faults with his work and making the section suffer a little for it so that 'we would not all be having this extra drill if it had not been for Bossy' becomes a feature of the general tenor of the section. This is a dangerous approach in that it may actually engender some sympathy for Bossy and resentment of Snow White if she is perceived to be picking on him. Alternatively, it may result in Bossy having an unpleasant experience at the hands of Psycho on a dark night when no one is watching, and that might just do the trick, but that is not generally considered good practice and Snow White would never countenance such a thing ... probably.

The social dynamic of the section can clearly be quite a challenge for Snow White to deal with and if there is any chance of the group being a viable unit on a day-to-day basis, she has to make it work. Unfortunately, this is only part of her professional challenge; the dynamic changes when the section goes into combat and she has to allocate individuals to tasks. From time to time, somebody has to be selected to move off in front of the section to locate the enemy or to ensure that a building or obstacle is not occupied. Who will she choose? For obvious reasons both Clumsy and Noisy are always going to be poor choices for a stealthy approach – and potentially Stinky is not much better. Psycho is always going to be happy to take the point position, but is more likely to do something rash if he sees the enemy and Lazy may well fail to notice an enemy presence or a trip wire until it is too late. In any case, what Snow White cannot do is continually give the job to the same one or two individuals. Even if they do not come to resent always being the man at the very sharp end they will eventually become demoralised and/or exhausted and may simply lose focus and lead the section into a very unhappy situation that could have been avoided.

In combat, of course, there is always the danger of being wounded or killed. The loss of any member of the section brings Snow White another set of problems. In the short term, she has to ensure that the section carries out its duties with fewer people, which of itself raises the risk factor for them all as well as having an impact on morale – even if the lost person or persons were unpopular, they were still a part of the group.

In due course the losses will be replaced and Snow White will have to embark on yet another challenge; that of integrating the new members into the section family. This might occur quite easily, but that is relatively rare. Newcomers are seldom welcomed immediately in any setting, and less so in one that is fraught with danger. It can happen that the newcomer or newcomers are distrusted professionally – they may have had the same training regimen as everyone else, but the section has accumulated some experience and are accustomed to one another. In extreme cases, new personnel can be regarded in a diffident or even hostile fashion; the replacement soldier is hardly thought of as an individual, but – to use a phrase current in the Vietnam war – as the FNG or 'F★★★ing New Guy'. Not 'Where's Joe' or 'Where's Fred' but 'Where's the FNG? Once again, the responsibility for dealing with a difficult problem in dangerous circumstances falls on the shoulders of Snow White.

The most remarkable thing about junior leaders like Snow White is not that they manage so well – which mostly they do – but that they can cope at all. Fortunately for all concerned, our heroine has had the benefit of a good deal of training to help and guide her and has also become familiar with her own role as a leader and tactician on the battlefield; the situation is that much harder for the person who is chosen to take her place if she has the misfortune to be wounded or killed. He might have the benefit of practical experience and have a good idea of the strengths and weaknesses of the section, but whether that will be enough to make him an effective replacement for Snow White is a different matter.

Making the section work as a team is clearly a difficult business, made more so by the fact that Snow White will have to give orders that will risk the lives of her colleagues; the whole process is a lot more demanding than anything that happens in a civilian workplace. The leader of a firefighting team may have to issue instructions that put lives in danger, but not as part of a twenty-four-hour day in which an opposing entity is making every effort to kill his or her workmates.

The same challenges – and a few extra ones just for good measure – apply to all combat arms. The man who makes the breakfasts in a warship is just as much part of the process of battle as the man who operates a weapon, pilots the vessel or monitors the radar images; furthermore, all of the issues that Snow White faces as a section leader are duplicated at every level of command. (A good example of a 'team player' might be the Beach Master on D-Day. Effectively, he was a traffic cop, a non-combatant. But controlling movement 'from the three-fathom mark to the high water mark' often made his job even more dangerous than that of the infantryman – and remember, he never got off the beach during the invasion.)

Twenty years after her stint as a junior leader in an infantry unit, we find that Snow White has made a good career in the army. Through exemplary service and assiduous study she has been commissioned and passed through Staff College with distinction. She performed well as a platoon and then company commander, in various administrative posts and as a battalion commander, and is now in charge of a regiment of three battalions and a number of attached specialist formations – reconnaissance, intelligence, nuclear and chemical warfare teams to name but a few. Much of her work consists of planning integrated operations which will be entrusted to the three battalion commanders; Wynken, Blynken and Nod. That's where her command problems start.

Wynken is by far the most proactive of the three. He is conscientious and intelligent, he has a good grasp of the general policies required, and he is careful of the lives of the troops under his command, but not to the extent that he shies away from difficult and dangerous projects. He also thinks that he should be in charge and he may even be right. Blynken is, all in all, a reasonably capable officer but prone to panic a little when matters take a turn for the unexpected and to lose focus when there is little activity on the front line. Nod, on the other hand, simply is not up to scratch.

Congenial, popular and well read in his subject, Nod has made a good career despite a woeful lack of talent. He drifts off during briefings and any verbal instruction has to be given twice … and not infrequently explained again in the middle of an operation. He went to the right school, has the right accent and has the right friends. With the best will in the world, Snow White cannot sack him, but equally she cannot depend on his battalion to take the leading role in combat operations. This is not a reflection on the unit, just on the commander. Since Nod is not and never will be a genuinely useful member of the team, Snow White is obliged to hand the most difficult problems to Wynken and the less challenging ones to Blynken. As a result, her own superiors have come to think of Nod as an effective officer. He has never failed in an operation and his casualties have been commendably low. It will probably come as no surprise to Snow White when Nod is given a regimental command and – who knows? – one day become her own superior as a Divisional commander. Part of the problem with Nod is that despite his tendency to switch off from time to time, he is generally a very active officer. Clausewitz made the observation that officers who are stupid and/or ineffective and idle are preferable to those who are stupid and/or ineffective but industrious. The industrious one does far more damage simply because he is so active. The idle one does very little at all so his impact is rather more limited.

Just to make matters a little worse, Nod may well be genuinely unaware that he is a waste of space. He's been promoted regularly so surely he must have been doing something right, in which case why should he change his behaviour? This is not, of course, a situation that is found in the military alone. Readers from a business background are very likely to recognise a 'Nod' in their own environment. If not, they may want to give some thought as to whether they are themselves the Nod-figure in their organisation.

6. What We Think We Know

Perhaps the most challenging – and generally the first – barrier to overcome when we start to study both the theory and practice of war is the raft of concepts that we might categorise as 'the things everyone knows'. All in all, these do not bear too much critical examination and some are simply untrue, so it is worth giving them some thought. Most of these concepts can be, and are, encapsulated in cheerful little proverbs, some of which we have grown up with and are ingrained by constant repetition, but that does not make them valid. Additionally, many military terms are used inappropriately or with a limited understanding of how they might be applied.

Two good examples of 'things everybody knows' but in reality are best avoided are 'God is on the side of the big battalions' and 'History is written by the winners'. Even the most cursory reading of one of the legion of books with titles like 'Great Battles of the World' or 'The Art of War Throughout History' should have made us aware that a great many battles and wars are won by the side with the smaller army and perhaps the smaller economy, but with a better motivated and/or better trained army. Most of the examples of victory and defeat that are well known would certainly suggest that God is actually on the side of the small battalions. Make a list of great victories and it might very well include Waterloo, Agincourt and Fredericksburg; all are examples of the smaller army defeating the larger. This is, of course, slightly misleading. If we follow the scholarly precepts of the great English historiographers Sellar and Yeatman we learn that history does not consist of what has happened but of what is *memorable*. A great many victories in war are memorable simply *because* the smaller force defeated the larger force.

An apparent disparity in size can be very misleading. A small army with excellent weapons and training will generally be a match for one with shoddy or obsolete equipment and little (or just less valid and relevant) training. Moreover, the face value of numbers as reported can be misleading. If a source gives the strength of a particular army at a given moment as being 20,000, do we necessarily know how many of those people were actually soldiers at all, let alone combat soldiers? If the figure was recorded on the first of the month as the army marched off to find the enemy, was it still valid a month later when battle was joined? In the intervening weeks, the army might have been radically reduced by desertion and disease, a portion of the army might have been detached for operations elsewhere or might even have been in a state of mutiny, and the student of the battle a month or a year or a century down the road would be none the wiser.

If God is not, after all, on the side of the big battalions, what observations can we make about the aforesaid winner who allegedly writes our history? In the wake of the First World War a great many volumes were published in English and French which described the process of the conflict, and a number on the development of practice, but really not all that many on how the war was won. In the same period, a considerable number of volumes were written in German about how the war had been lost and about who was to blame. Essentially, the winners do not need to explain victory – they won, simple as that – whereas the vanquished often feel a need to explain and thereby excuse their defeat. That may take the form of scholarly and detailed analysis (which may or may not be valid) or it may be more a matter of writing about specific successful battles or very courageous individuals or about commanders and armies that would have been successful had they been given the sort of political, economic or professional support they deserved. A brief survey of the enormous range of popular works available on the Vietnam War would provide any number of good examples of this phenomenon.

The same sort of thinking that underlies 'big battalions' is at the root of the easy certainty that a 'better' army guarantees success. It can be applied – rather romantically – to particular institutions or weapons or tactical practices. If the Germans had such wonderful soldiers and weaponry in 1939–45, for example, the Waffen SS, the King Tiger tank, the MG 42 machine gun, the blitzkrieg concepts, then we have to wonder how they had not won the war by the end of 1942. There was, of course, a vast range of other factors in play but the effectiveness of a specific weapon, formation or battlefield drill is seldom, if ever, the sole agent of either defeat or victory.

Another favoured expression – and one that is strongly indicative of a profound failure to understand the subject – is 'cannon fodder'. The simple reality is that commanders never have spare soldiers; in fact, they very seldom feel that they have anything like enough. They certainly never set about a project with the intention of getting a lot of soldiers killed to no useful purpose. That may be the eventual effective outcome, but it is never the premise of an operation.

It could be argued deploying a modest force as bait to persuade the enemy to commit a large force to a project that looks like it will produce a positive result, but in fact is no more than a trap, is tantamount to the same thing. That is far from the case even when it entails the likely loss of the unit or detachment. Successful entrapment is a matter of persuading the enemy to use up his resources in a futile endeavour, but if that is to be achieved, the enemy has to believe that there will be a benefit which will outweigh any losses because he – like commanders everywhere – does not believe in the cannon-fodder concept either. In fact, the commander who deploys the 'bait' must believe that the loss of manpower will make a positive difference to the course of the battle without causing serious damage to his own force. His intention should be – and generally will be – actually to reduce his own losses over the course of the operation by inflicting a heavy blow on the opposition before the opposition can inflict a heavy blow on him.

Probably the most popular of all the old saws is 'military intelligence is a contradiction in terms'. It is certainly true that military decisions have to be made in the absence of all the information that the commander would like to have, and it is not uncommon that a decision made in the light of very limited information can be wrong. This is the case in any walk of life. Businesses and institutions make poor, bad or even dreadful decisions all the time. One could even be forgiven for thinking that the political class is actually dedicated to making poor decisions. The difference is that when a commander makes a poor decision it tends to involve loss of life and limb – sometimes on a very large scale. This occurs despite the very best efforts of very fine minds that have been trained for analysis and equipped with a huge knowledge of past operations stretching back thousands of years; it happens despite the intelligence effort, not because of it. It is, however, an easy joke to make and it has enjoyed a steady currency for a long time. That does not make it valid, just a commonplace.

Not all catchphrases are invalid. The phrase attributed to various commanders of the American Civil War that success in battle was a matter of getting 'there firstest with the mostest' has a certain ring of veracity – indeed, similar sentiments have been expressed by most if not all of the great military thinkers. It is worth thinking about what it really means and what is significant in the phrase. The most important issue is not numerical strength (mostest) or even speed of concentration (firstest) but location of effort or 'there'. Unless the commander can identify where the forces need to be, there is little value in having them concentrated in great numbers or in great haste. Unless he is very fortunate, he will just end up with a lot of mouths to feed in a place where they really do not need to be.

That is even worse than it sounds. If the forces are not in the right location they must effectively be in the wrong location. Naturally, that is a bad thing for the commander, but it is also a good thing for his opponent. Your enemy may not be in the optimum location either, of course, but if they are not then it is even more important that you are. If you are not, then there is little chance of a positive opportunity arising and accordingly little chance of achieving a victory, which is, of course, what commanders are generally there to do. There is a caveat in the use of 'generally' and 'victory'. Success is not simply a matter of seeking victory on the battlefield. It can be achieved by avoiding combat and, on the whole, is almost invariably helped by avoiding unnecessary engagements. Clearly, fighting an unnecessary battle and losing is always a bad thing and is a failure of command twice over – the commander fought when he did not have to and then compounded the error by losing as well. To make matters more complicated, fighting an unnecessary action may be disadvantageous in a number of ways even if the fight is won.

Obviously, there is a cost in people and material which may not be easy or even possible to replace. The force has been weakened to no useful purpose and may now be vulnerable to a counter-attack by a reorganised opponent, who may even be better motivated than he was before due to a desire for vengeance.

There are other risks. The temptation to inflict a defeat on an unwise or incompetent opponent may lead to that opponent being dismissed and replaced with a more competent adversary. Either of these possibilities could be summarised as 'never kick a man when he's down … he may get up and hit you'.

Naturally, a victory – even one that is broadly speaking an irrelevance – is likely to have a positive influence on the morale of the man in the field and at home as well as a corresponding effect on the morale of the opposition, but it might just as easily have the unintended consequence of reinforcing the enemy's will to fight – the attitude of the British after the defeat of 1940 is not unique.

Battlefield success might also have an undue influence on the policy-makers. If a successful battle results in the acquisition of territory – which it very often does – and even if possession of that territory is irrelevant or even detrimental to the overall strategy of the campaign – it becomes a hostage to fortune in several ways. The senior staff or the political establishment may come to see giving up that territory as a defeat in itself, and the troops may think that failing to hold the territory means that the sacrifice of their comrades has been in vain. Gaining ground is one of the factors in war that is most easily understood in terms of failure or success and there are reasons why that should be the case, but it is not universally sound. Even if a battle has been fought successfully for cogent reasons, retaining real estate may not be a desirable outcome. More land – broadly speaking – means more soldiers are required to hold on to it. If reinforcements are not forthcoming (or at least not in a useful time-frame) the army – probably weakened by losses in combat – now has to cover more ground with fewer troops. The enemy might, of course, have been sufficiently weakened that this is not an immediate concern, but if that is not the case the victors might have become more vulnerable to a counter-attack. The situation might be worse yet. The ground gained might not be as suitable for the victors as the terrain in which they were operating previously. A victory that takes a mobile armoured force into dense forest or urban areas might be very costly indeed in the longer term.

A battlefield success that carries the army into country where it will be harder to supply might simply lead to a starving or immobilised force, but there might not be a better alternative. If the enemy is defeated in the field it might be necessary to pursue him – regardless of the nature of the terrain – to prevent his recovery, to provide an approach to more distant objectives and/or contribute to the wider strategy of his military superiors and/or to fulfil the policy goals of the political establishment.

Winning battles is therefore a tricky business and fraught with challenges and risks. Losing battles is even worse, but it is the lot of – roughly speaking – half of all battlefield commanders. It is all too easy to criticise defeated generals; they should have done this or that and they should not have done that or this. This sort of criticism is sometimes well intentioned, sometimes valid, sometimes ignorant and sometimes simply dishonest. It is most generally delivered by people with little or no understanding of the structures of the forces, the tactical or strategic environment or any thought as to what information was available to the defeated commander, the rationale behind his plans, his instructions from above or whether his subordinates carried out their instructions effectively – or at all. It is also mostly delivered from the comfort and security of an armchair in an office or study or in front of a television camera. That does not mean that the criticism is necessarily invalid, but it is seldom genuinely well informed and may be extremely unjust.

Historians and journalists are inexcusably prone to ill-judged conclusions, often because they have an ideological position which they feel justifies a certain analysis or because they have simply failed to grasp the realities of the environment in which this or that general had to make decisions, but very often it is no more than the time-honoured practice of being wise after the event. With the benefit of hindsight, this or that aspect of a situation is 'totally obvious' to the writer ... though if it were quite so obvious at the time, one imagines that the commander might have noticed and taken action accordingly.

That is not always the case. Military commanders can be just as incompetent as anyone else and there is no need to look far for examples. Soldiers who have the right connections (people like 'Nod' in the last chapter) or just 'talk a good show' can rise to senior rank despite a lack of ability, but since the same holds true in business, and is almost universal in political life, perhaps we should not seek to hold the military to a higher standard. If we cannot prevent greedy, self-obsessed idiots from getting to the top in political life, we should hardly be surprised if they sometimes get to the top in the military.

A misleading aspect to our perception of defeat is the belief that commanders – or, for that matter, the entire military establishments of nations – who have been defeated on several occasions have simply repeated a formula that has already failed them in the past and that they have done so through a mixture of stupidity and arrogance or a total lack of imagination. This can be true; it can also be true of commanders who never lost a battle in the same way twice. Once we start to look at a battle or campaign in more detail, it is more often than not the case that what at first seems to be repetition of a particular approach is not so.

Looking at the struggles on the Western Front in the 1914–18 war could lead us to the conclusion that each offensive by either the Germans or the Allies conformed to the same principles and failed for the same reasons. In fact, several distinct variations on a similar theme were utilised, but the options were severely restricted by the environment. Once the pattern of two opposing lines of fortification had been established, there was no other course beyond frontal attacks if the political imperatives were to be attempted, let alone achieved. As a rule, the preferable course of action in dealing with a strongpoint must be to neutralise it by passing around a flank and isolating the position from its chain of supply and command. Since the trench systems stretched from Switzerland to the sea there were no flanks to exploit and the entire front became, in essence, two gigantically long strongpoints facing east for the Allies and west for the Germans. Numerous novel approaches – and some old favourites – were applied to the problem, but none was equal to the task. Immense bombardments were applied to destroy the belts of wire which prevented easy movement of infantry, only to discover that it is very difficult indeed to destroy wire entanglements with high explosive. Even heavier barrages had little more effect and increasingly intense barrages lasting, eventually, for days on end did not make much of an impression either.

With the benefit of that special wisdom entrusted to casual observers after the event, all that is 'abundantly obvious', but it would not have become so if the attempt had not been made. Gas attacks had some utility, but were dependent on having a breeze – but not too much of one – that would blow consistently in the right direction for a long enough period, but in any case the introduction of reasonably effective gas masks prevented such attacks from becoming a standard operational practice. The terrain – especially in Flanders – became increasingly difficult to deal with through the effects of combat. Massive shelling did not just churn up the topsoil of the landscape; it destroyed the agricultural drainage systems that had made the land so fertile and thus prevented the escape of surface water, so the ground between and around the front lines became and remained utterly waterlogged, a veritable quagmire. Clearly, that was a factor in preventing successful infantry attacks, so steps were taken to counteract the problem of the mud ... the British invented the tank.

Contrary to 'perceived history', Field Marshal Haig was not implacably opposed to innovation and grasped the potential of the tank at once. He did not, however, want to use the new platform immediately. He wanted to wait until the tanks were available in large numbers so that they could be committed en masse before the Germans developed any useful countermeasures. His opinion that this would be the better course and would

German gas mask training; it is estimated that over 90,000 were killed by gas attacks in the First World War and as many as 1,200,000 were affected to some degree. But it was not a war- or even battle-winning weapon.

provide the breakthrough to open country that both sides were anxious to achieve might well have been valid, but he was over-ruled by Lloyd George, who felt that war was too important to be 'left to the generals'. Lloyd George insisted that the tanks be put into action before they could be supplied in large enough numbers to make a major breakthrough, so a huge potential tactical advantage was thrown away, since the Germans immediately applied themselves to developing countermeasures such as the anti-tank rifle.

The commanders on both sides were faced with problems that had not arisen in the past and, tragically, in war there is no means of knowing whether an innovation will work until it is tried. It is not altogether unfair to assert that the generals had – mostly at least – failed to grasp the significance of all of the military and social developments of the preceding years, but there again neither had anyone else. The French and the Germans at least had the experience of organising huge forces – though whether they did it well or badly is a different question – but only for very short periods. Universal conscription in both countries meant that there was a massive pool of men who had undergone military training, and the French and German governments and military administrations had a better grasp of the scale of material and transport required, but neither had envisaged a war involving millions being conducted for years on end.

The British had not really considered such a prospect at all. The army of King George was positively tiny by European standards and had been designed – rather well as it happens – for a very different range of challenges. The men who commanded the British forces in 1914–18 were not unusually stupid or ill educated. In fact, most of them had studied and trained for war throughout their adult lives, but they had been preparing for the sort of conflict that an imperial power might well face at any time – insurrection in colonies or incursions by third parties in any corner of the world, not a mass conflict in France and Flanders.

It is legitimate to ask why they had not trained for war in Europe, and the answer is quite simple; they had not been briefed to do so, and even if they had been so inclined no British government would have been willing to foot the bill for the training and procurement of necessary equipment or supplies. The political class was convinced that home defence could be maintained by the power of the Royal Navy – the most powerful maritime force in the world – leaving the army to deal with problems further afield. The remarkable thing really is that the British Expeditionary Force was not utterly destroyed in the first few weeks of the war. This is a testament to the quality of training and leadership of the army as a whole, but also to the remarkable capacity of senior and junior officers to adapt to a situation so far removed from their *raison d'être*.

Both sides had to cope with conditions that could not have been foreseen and both sides had to do so with the assets that were available. Soldiers can only fight the battle in front of them and they do so on terms that are set by the political establishment, generally with little or no regard for the environment in which the military will be deployed. This is inevitable to some degree; one should bear in mind that the political class is not equipped with crystal balls – they cannot always predict the threats that might arise, but tragic mismatches occur all too often and in virtually every country despite the lessons that should be, but often are not, learned from history.

Those lessons include the failure to appreciate that the enemy might have learned something from their own experience and that of others. Hitler's invasion of the Soviet Union achieved incredible penetration in the summer of 1941, but failed to achieve the knock-out blow that would have put the Soviets out of the war. The deep penetration offensives in Poland and then in France, Belgium and the Netherlands had been immensely successful, but the sheer scale of the Russian landscape was a match for that technique. Stalin's armies could retreat for hundreds of miles to avoid battles of annihilation, but the western Allies simply ran out of space and found themselves with their backs to the Channel and no choice other than to abandon France. When winter came, the Germans did not have the clothing or the equipment necessary to operate in the cold weather – vehicles were immobilised because the fuel froze in their tanks and troops froze to death.

The experience of Napoleon's invasion of 1812 should have made Hitler aware of the risks, but he chose to disregard the warning. He doubtless believed that the situation was not the same; after all, Napoleon had no trains, trucks or aircraft, but the corollary is that Napoleon did not have to procure petrol and lubricants for his tanks and planes or cope with the enormous consumption of ammunition that is so crucial to modern warfare. The situation was not the same, but the principles were not different, and any of Hitler's generals could have told him so ... but only at great personal risk and to what end? A politician seldom abandons a plan simply because it is bad, and in that respect Hitler was not particularly unusual.

An apparent failure to learn from previous experience is often just that, apparent rather than real. Commanders do not set about their obligations with the intention of defeat, however often it may seem to the casual observer that they have done exactly that. Counterinsurgency is, perhaps, the most obvious form, though to a degree that is itself driven by cultural and social expectations and by a somewhat romantic view of the genre. There is a certain attraction to the triumph of the weak over the powerful. That relationship can be misleading and it is not uncommon for one side to use their underdog status as a propaganda tool in the struggle.

Looking at the Vietnam War, it is very easy to absorb a view that is popular but not especially valid. The conflict is widely seen as a struggle between North Vietnam and the United States rather than one between North Vietnam, South Vietnam and a number of allied nations and supporters on both sides. South Vietnam had the support of Australian and South Korean troops as well as the economic and military aid of the United States. North Vietnam received aid from the Soviet Union and her satellites as well as from China, and, of course, the conflicts in Cambodia and Laos had a considerable impact as well. From the popular view of the conflict as a fight between a small nation – North Vietnam – and a superpower in the shape of the USA, it is absolutely clear that America had the upper hand in terms of economic and military power, but the equation is not so simple as all that. On a strategic level we might equally see the conflict as one between two of the biggest armies in the world. The USA had a very much larger population, but even with conscription only a relatively small portion of her people could be mobilised for war compared to the 'total war' commitment that was applied in North Vietnam.

For a variety of political and social reasons, the American conscripts could not be retained in service for the duration, as had been the case in the Second World War. Effectively, the enlistees were posted to Vietnam for a year or so. By the time they had adjusted to the climate and gained enough practical experience to be valuable on the battlefield they were, naturally, much more likely to be focused on the end of their posting and going home alive than on applying themselves to the job in hand.

Additionally, though the North Vietnamese army was, of course, very much the smaller of the two, the Americans had to furnish troops to fulfil commitments all over the world. Although the enormous American commitment in Europe was focused on the possibility of a conflict with the Warsaw Pact nations in Germany, we would be very much mistaken if we failed to take account of the Cold War in Europe as an aspect – however coincidental – of Soviet support for North Vietnam. American troops were needed, or so it was believed, in any number of locations, whereas North Vietnams troops were – almost without exception – only required for roles in Vietnam itself. In fact, due to the political and diplomatic policy position of the United States, they were, in the main, available for service at the front in South Vietnam since there was no real prospect of North Vietnam being invaded. We should, of course, consider all sorts of other factors – primarily the unpopularity of conscription in general and the Vietnam War in particular – in American society as a whole, especially after the Tet offensive of 1968.

The fact that the Tet offensive was actually an unmitigated failure for North Vietnam was neither here nor there; it seemed to the public that years of conflict had not 'turned the tide against Communism' in South East Asia, and support for the war dwindled rapidly. For the rest of the world, the Tet offensive could be seen as an indication that God was not 'on the side of the big battalions' after all. This was not wrong, but the horse had been put before the cart. The Army of the Republic of Vietnam, the Americans and the other allies were, in fact, outnumbered, and despite being attacked during a truce, still secured victory at both a tactical and strategic level in remarkably short order. It just did not look like that to the wider public, either domestically in the USA or South Vietnam, or across the world as a whole.

In North Vietnam, the Tet offensive was, naturally, hailed as a great success, but in a society where the government has total and complete control of the media that is not a hard case to present. If nobody is allowed to offer critical analysis, the populace is, by and large, likely to believe whatever it is told. Whether it is true or not is hardly really a factor at all.

Sticking with the Vietnam War for a little longer, we might consider the general policy of the Americans and their allies as an example of commanders repeating the mistakes of the past, both tactically and strategically. What we should not do – assuming that this was actually the case – is accept that that is what they believed they were doing. The strategy of erecting powerful defended bases and communities and thus preventing the free operation of the enemy was not invalid. It had worked perfectly well in other theatres and at other times, but it had very clearly not worked for the French in the 1950s, so why would anyone expect it would work for the Americans in the 1960s and 1970s?

We should assume that the commanders were either incredibly stupid or that they did not see the conditions as being identical. The resources of the French forces had been very much less than those of the United States, especially in airpower, both as a strike weapon and as transport. The Americans would have been able to deploy much larger numbers of airmobile troops to flashpoints, and overall would have been able to deploy larger numbers of men and greater quantities of armour and artillery – and of course the French had also been fighting a war in Algeria at the same time. The political situation was also rather different. The French had been fighting a war to preserve a colonial regime, whereas the Americans – ostensibly – would be fighting to support an allegedly democratic government of an independent country; they were not imposing an alien administration, but preserving a domestic one against a foreign totalitarian power, not to mention 'striking a blow against the threat of communism'.

Seen in that light – and that is not to say that the analysis was valid or that the policy direction was viable – American intervention in Vietnam could be seen as little or no different from the British going to war against Germany in 1939. It had proved impossible to curtail Hitler's ambitions by either negotiation or appeasement, so war became the only option. In both cases, policy adopted by the military might seem to have been at best flawed and at worst totally irrational.

Britain and France declared war, but did not actually pursue any kind of practical offensive strategy. Various well-worn military phrases can be applied here, and, for a change, they are not inappropriate. 'To defend everything is to defend nothing', and 'the inevitable outcome of a defensive war is defeat'; both fit the bill reasonably well. For political and social reasons – not entirely different from those of the Americans in the 1960s – the French and British political establishments decided that they would not mount offensives against Germany, but would, instead, rely on 'impregnable' fortifications that would stop the Germans in their tracks if and when they chose to attack.

The experience of the First World War gave a superficial credibility to the policy in that it was all too clear in the 1914–18 conflict that strong entrenchments backed up by well-organised artillery in large quantities could inflict unacceptable losses on an attacker … a doctrine of deterrence. This did not, as many writers have claimed, ignore the development of the tank; quite the reverse, in fact. The Maginot Line installations included all the measures which were – as was understood at the time – necessary to prevent armoured incursions. Neither Britain nor France was prepared to countenance an attack on Germany for fear of a repeat of the enormous losses of a generation before. The policy was clearly not entirely irrational; had they seen it in that light they would, one imagines, have adopted a different one. The option of mounting an invasion was not, however, as good a prospect as we might expect. It was certainly the case that in the late summer and autumn of 1939 the bulk of German forces were still in the east and that transporting them to the western front would have been a major challenge, but the Germans had their own version of the Maginot defences; the Westwall or, as the Allies called it, the Siegfried Line.

How effective the Westwall would have proved is open to question, but there were other issues to consider. Invasion of Germany would have involved a risk of losing diplomatic sympathy abroad and would not necessarily have played well at home. France had suffered far more heavily than Britain in the First World War and the prospect of hundreds of thousands – perhaps millions – more deaths was not something that would play well among the electorate; this was also true in Britain. Moreover, there were strong far-right movements in both countries which were not unsympathetic to the rise of National Socialism.

If the political situation was difficult, the military one was no less so. After the armistice of 1919, the British political parties had moved rapidly to their traditional policy position of having a very small army, focusing more on an extensive navy which could, or so it was claimed, protect the country from invasion. There were a number of faults with this line of thinking, not least that the navy was required to patrol shipping routes and protect colonial possessions all over the world. Additionally, the burden of debt incurred as a result of the First World War, and the economic trials of the 1920s and 1930s, meant that there was simply not enough money to sustain an adequate army, and the death toll of 1914–18 had – for perfectly understandable reasons – made military investment deeply unpopular, so even if the money had been available, the political will was not.

Consequently, Britain's army was not large enough to sustain a major offensive in Europe and was not really equipped to do so. Nor was it trained for such a function. The condition of the army was, in a sense, similar to what it had been in 1914. It was an army designed to support locally enlisted forces in an imperial environment. The French did have a large army and, having continued with conscription after 1918, there was a very large pool of reserve forces, but the army was not well designed for offensive purposes; it had been assumed that the Maginot Line might be penetrated and that there might have to be counteroffensives, but the general thinking was reactive rather than proactive. An obsession with defensive measures had prevented the French – and the British too – from developing an offensive doctrine. However strong and well designed the *Gros Ouvrages* (major installations) of the Maginot Line might have been, they were still vulnerable to heavy bombardments and – unlike in 1914–18 – the Germans were not obliged to maintain an extensive front; they could concentrate their efforts on selected locations. In fact, they did not have to take much action against the Maginot Line because they could go around it. For diplomatic and political reasons, the line did not extend along the Belgian border, and efforts in 1939–40 to correct this weakness were less than effective.

The defeat of France and Britain in 1940 can be – often is – laid at the feet of the generals. They were unimaginative, they were conservative, and they had failed to absorb the lessons of new techniques and technology. There is some truth to this, but they were also constrained by the policies of their governments and were put into impossible positions. Just because the British army was not designed for a European conflict – just as in 1914 – did not stop the British government from committing it to a war on the continent. It is of course true that any task can only be approached with the tools and techniques that are available, but the French and British governments had chosen the tools, not the commanders.

The political class may have responsibility for the setting that commanders have to work in and the material that they have to work with, but there are factors beyond the control of governments, such as an unexpected political crisis or economic collapse, and they are also sometimes constrained by the needs or expectations of the community. A particular course of action may be demanded or prevented by public opinion.

The massive bombing campaign against German industry and cities in the Second World War was not effective, and the British government became aware of this quite early on in the process, but could not abandon it. The cost in men, aircraft and industrial demand was very heavy and the resources could have been put to better use, but once the process had started it acquired a certain social and political momentum of its own. Ending the campaign would have been seen either as an admission that the government had adopted a poor strategy or as an outright defeat, neither of which would have been good for confidence. Additionally, it could be seen as a slap in face for the friends and relatives of the thousands who had been killed and the men who had been maimed or shot down and captured. Worst of all, the bombing strategy was perceived by many as a means of hitting back at the enemy and it had certainly been presented to the public by newspapers, radio and newsreels in cinemas as a serious blow to the Nazi war effort. Either the public was to believe that this was the case, in which case why bring it to a close, or else the bombing had not been a success at all, in which case the government and the media had simply been lying for some considerable time. None of these was an acceptable outcome for the government, so the campaign continued, because the government really could not take the risk of giving up on it.

That is a very specific example, but, more generally, the public does influence defence thinking, even in a country like Britain where – all in all – there is very little interaction between the military and the civil community. The influence may not have any basis in practicality. Every time a government conducts a 'defence review' and proposes disbanding one or other of the 'historic' infantry regiments there is something of an outcry. This was not always the case and the outcry may be short-lived, but it can be successful, even if only on a very superficial level. A popular campaign to 'Save the Argylls' (a famous infantry regiment) in the late 1960s achieved its objective, but the government of the day simply made cuts elsewhere. Whether the cuts were rational or not, and whether they were a better or worse option than disbanding one infantry regiment, is a different matter, but there is never any public concern if a government proposes to amalgamate two support-arm departments; no one campaigned under a banner to 'Save the RASC' (Royal Army Service Corps) when it was combined with other departments to form the Royal Logistics Corps.

Public perception can have a long-term effect on matters which neither they nor the political class actually understand. Any plan to disband the British Parachute Regiment would be met with a great deal of hostility because the 'Red Berets' are so famous, so famous, in fact, that hardly anyone remembers that the berets in question are in fact maroon. More importantly, no consideration is given to the fact that the fame of the regiment is largely based on its near-annihilation at Arnhem in 1944 or to the fact that having a parachute formation is of limited value if there is an inadequate supply of aircraft to transport them.

Phenomena such as public support for the Parachute Regiment are not simply a matter of fame and romance, though there is an obvious connection. It is also a product of perceptions shared across the political class and within the military itself and, once again, there is a trite and treasured phrase; the army (or navy or air force or government – delete as preferred) works to 'prepare itself for the last war, not the next one'.

There is just about enough truth in that to make it worth its existence, but no more than that. Any organisation is the product of its experience. The men and women who conduct the business and plan for the future have – hopefully – been trained and exposed to a variety of experiences that will prepare them for the task. Inevitably they – and the people who select them for promotion, design their training plan, purchase their equipment – must be guided by the things that they have seen and the studies they have undertaken in the course of their own career. It would be remarkable if they did not choose to promote men and women with very similar outlooks. To a great extent, planning for the last war – or more accurately, planning on the basis of the experience gained in the last war – is inevitable and not entirely undesirable. If the last war resulted in defeat it is not unreasonable to conclude that there was something wrong about one or more aspects of the approach taken, and if it resulted in victory it is not unreasonable to conclude that there was something right.

Either may well be the case, but neither is a truly reliable guide to future development. No two conflicts are really the same and the maxim that 'history repeats itself' is simply not true. The approach that brought success or failure in one situation may not deliver the same result in another. The lessons learned in North

Africa in 1942 may have had some value in France in 1944, but in some ways were actually detrimental.

Experience – whether of defeat or victory – conditions the thinking and the practices of the soldier at all levels. In 1914, a British Warrant Officer (senior NCO) took it upon himself to organise the unit's transport into what he called a 'Zariba' – a defensive enclosure rather like the circle of wagons which feature in any number of 1950s westerns. His experience and knowledge had taught him that this was a valuable thing to do, but that experience was derived from fighting against ill-equipped opposition in colonial campaigns; it served no useful function in the very different environment of France. In fact, it just provided the enemy with a concentrated target for machine gun and artillery fire. The warrant officer in question was not stupid or lacking in motivation; he was a victim (for want of a better term) of the risks of doctrine.

Armies cannot function without doctrines relating to defence, offence, logistics and everything else. Bluntly, you have to start somewhere and where better than with a doctrine to which the army is accustomed and which has some track record of effectiveness? Large organisations – not just armies – need to have general practices to enable one part to relate to another in a manner that is comprehensible to every part of the whole; the problem arises when the doctrine becomes doctrinaire and when the enemy has become sufficiently familiar with that doctrine to develop one of his own which is superior. This, of course, is why soldiers should – and generally do – study examples from history. It is not the case that extensive knowledge of past campaigns will enable one to identify identical situations and come up with ideal – or at least successful – solutions to every problem, but at least it can provide them with clues and insights which can – hopefully – help them develop an effective tactical or strategic approach to the problems that they face in the here and now.

The challenge lies in knowing when to discard the doctrines that have succeeded but might no longer be valid. The American Civil War was largely conducted in the light of the tactical approaches of the Napoleonic Wars half a century earlier. Bodies of infantry two or three ranks deep deployed in line and exchanged fire until one side or the other had had enough and withdrew or until one side made a rapid advance toward the enemy to force a hand-to-hand fight. That was the general idea, but in practice it seldom really materialised. Either the attackers would lose their determination to come to blows, grind to a halt and then retire, or the defenders would lose heart and not hang around to see what might happen as and when the attackers made contact.

Received history would have us believe that the retreat of the defenders would result in heavy casualties during a vigorous pursuit by the attackers. In practice, once the defenders had vacated their position the attackers would be much more inclined to halt there and then pursue their opponents with nothing more deadly than shouted insults. A pursuit might be preferable, and officers would encourage their men to press on, but other factors came into play. The men who had made the charge would be somewhat tired or at least reacting to the receding adrenalin reaction brought on by the stress of the advance. They might prefer to savour their moment of victory, and they might also prefer to watch the enemy run rather than take the risk that they might regroup and fight back.

There are, of course, many examples of successful attacks being pressed to a conclusion, but not so many as one might think. All the same, steady fire followed by a charge had been the practice that had served well against Napoleon's forces in the Peninsular War and it would certainly be wrong to say that it did not serve effectively in the Civil War – so much so that the dense-column attacking style of the French was seldom if ever applied; the reliance on small-arms fire delivered from a relatively long, thin line was almost universal.

The tactics of 1812 were not actually unsuccessful in 1862, but the firefight combats of individual units tended to be shorter and the casualty rates were simply appalling, so what was different? Superficially, the weapons were of the same order in both conflicts; muzzle-loading muskets wielded by long, thin lines of soldiers standing shoulder-to-shoulder, but there had been a major technological development. Although smooth-bore muskets were retained by some units right up to the end of the Civil War, most men were armed with rifles and increasingly they no longer fired the round musket ball but a conoidal bullet. These new Minié bullets had a higher muzzle velocity and a much more reliable trajectory. They were more accurate over a longer range and the wounds inflicted were more severe. Clearly, there had been a major alteration in the dynamic of the firefight but not in the approach to combat, so we have to ask why not.

To some extent, the implications of a better weapon simply had not been appreciated by officers on either side. Both armies had officers who had fought in a major war; they had 'cut their teeth' in the Mexican War twenty years before and they had all been conditioned by victory there. The practices of the 1840s brought the same outcomes in the 1860s because both sides had retained them, but were there other options? By the end of the war, the shoulder-to-shoulder lines had not completely been abandoned, but they were employed less often. From the outset of the conflict it was an accepted practice to have a thin screen of skirmishers positioned in front of the main line of the unit or units to disrupt the enemy. This was not an innovation; European armies had been doing it with small bodies detached from units for decades, but by the end of the war it was not uncommon for a whole regiment or even a brigade to be deployed as a skirmish line, with a consequent reduction in casualty rates.

If this was an effective use of manpower in 1865, why was it not applied at the start of the war in 1861? There are several reasons and conservatism and poor judgement on the part of commanders were certainly factors, but not, perhaps, the overriding ones. No previous conflict had seen the use of the rifled musket on such a large scale, so in the early stages there was no appropriate experience to guide commanders. Moreover, although the armies became very large very quickly, the pre-war army had been very small indeed, only about 16,000 men, most of them scattered around the country in tiny garrisons. A very large proportion of the professional soldiers had seldom if ever had to control more than a company of perhaps eighty or ninety men and even fewer had done so in combat since the Mexican War. Even if there had been a wider understanding of the new weaponry factors, there was no time to develop a new training doctrine and then instil it into the officers and NCOs.

Beyond that, training men for war often has some connection to what they already know – or think they know – about the subject. The volunteers of 1861 would only have been aware of battle from what they had read or what their grandfathers had told them about the War of 1812. The general public understanding of battle was that it consisted of men in lines firing muskets at one another. New recruits today would have a similar sort of conditioning. They have seen scenes of conflict on the television and, for them, war is a business of tanks prowling across the countryside, aircraft firing missiles into cities and men with scarves firing AK-47s indiscriminately, possibly holding the weapon over their heads and shooting blindly from a window or from behind a pick-up truck. That perception is not incorrect in the sense that the news footage they see on television does reflect an aspect of modern conflict; a great deal of modern warfare does involve militia forces with limited training. That said, a not-inconsiderable portion of the footage on our screens is not really recording combat, just armed men firing their weapons for the benefit of the cameras.

7. Tradition, Change and the Perception of Success

Failure on the battlefield may or may not be a result of incompetence, and sometimes it is very hard or even impossible to tell. An officer who has been successful in a particular environment or at a certain level of command may not do well elsewhere, but there is no way of really ascertaining that other than giving him the opportunity to sink or swim. Past conduct and experience are not necessarily a good guide. People who perform admirably as junior and mid-ranking officers or even as staff experts do not always make effective senior commanders. The reverse can also be true. Men who do not have the personal attributes required of really successful platoon and company commanders might have the insight and determination that makes great generals, though they will seldom get the opportunity.

Defeat can be a catalyst for change, though often the alterations that need to be made are either not recognised at all or are not implemented adequately. Alternatively, the need to adjust the attitudes or equipment required for success may have been identified correctly, but resisted. The resistance may have been irrational, but sometimes the irrationality is more apparent than real and it is all too easy to attribute what seems to be a pointless adherence to the principles of the past to reasons that do not bear detailed examination. Many commentators on the First World War have been very clear in their belief that it should have been abundantly obvious that the era of the horse in warfare was long past. They were not altogether wrong, but they were not right either. The age of the headlong cavalry charge certainly should have been well and truly over – though there are isolated examples of successful cavalry actions in the classical mode right up to 1940.

The advent of accurate rifle fire and the machine gun did not mean that the horse had outlived its usefulness. Mechanised transport was still in its infancy. Most artillery still had to be drawn by horse teams and although ammunition and provisions were largely carried to supply depots by train, stores had to be distributed to units. This could have been done with trucks, but there simply were not that many trucks to be had and not all that many drivers. Horses and wagon drivers were available and were therefore put to use. Once the deadlock of the trenches was broken, cavalry became useful once again. They did not make sweeping advances across open country in quite the manner envisaged by some generals and politicians – or indeed quite a lot of the public at home – but speed and manoeuvrability meant that they could carry out effective reconnaissance or operate in a 'mounted infantry' role, advancing to secure a particular location such as a bridge or railway junction, dismounting and holding the objective until relieved by the main body of the army.

In principle, this was no different from the application of airborne forces in more recent years. The practice was not actually new. The reverses suffered by the British in the Boer war had led to sending more than 30,000 men to South Africa to serve as mounted infantry in the Imperial Yeomanry.

Although the British and German cavalry had made very little use of their lances in the cavalry actions of 1914 – they had quickly discovered that dismounting to use their rifles was a much better idea – the relative success of cavalry units in the great offensive of 1918 gave a new lease of life to the cavalry. Some of that success had been carefully fostered and assiduously exaggerated by soldiers who just happened to like horses and also by writers and politicians outside the military who were obsessed with the romance and excitement of the charge. All the same, the writing was on the wall and at the end of 1928 the British army started the process of mechanisation. It was not done with any great haste; the Scottish Horse did not lose their horses until mobilisation in 1939 and the Royal Scots Greys were still using their horses on operations until 1941.

Even then, the age of the horse was not at an end. The Russians and Poles – among others – still saw the mounted soldier as a useful asset and a whole SS cavalry division was formed in 1942. Mounted units were not limited to the Eastern Front; some American Reconnaissance troops in Italy took to horseback in 1943, the US Army 26th Cavalry conducted very successful actions against the Japanese at Binalonan and Bataan and even that most enthusiastic proponent of the armoured force, George Patton, believed that a cavalry division with pack artillery would have been a tremendously useful asset in the difficult terrain of Tunisia and Sicily.

Although the machine gun had made the cavalry charge redundant, the horse could, in certain circumstances, continue to be a useful military asset. The German army that performed the remarkable Blitzkrieg offensives might have relied on armour and aircraft at the cutting edge of the strike, but the majority of the army continued to be reliant on horse transport right up to the end of the greatest war in history.

For every proponent of the horse in the 1930s there was another 'expert' arguing that reliance on mechanisation was long overdue. Similar claims have been made in relation to other forms of military asset. The Yom Kippur war of 1973 allegedly involved very heavy losses in armoured vehicles to missiles that could be carried and operated by infantry. Almost immediately a host of writers predicted – even demanded – the immediate retirement of the tank. It was clear to them that the infantryman – at long last – could be provided with a weapon that gave him the edge over a 50-ton steel adversary. In fact, the overwhelming majority of armoured vehicles lost on both sides were destroyed

Horses before Sevastopol – though not at the 1854 siege but on New Year's Day 1942. The horses look even more miserable than the men. (From Paul Botting and Vince Milano's *The Lost Landsers 1941–42*, The History Press)

by aircraft or by other tanks, but that did not make so dramatic a tale.

This does not mean that the role of the tank is in some sense sacrosanct, and it is easy to overrate its invulnerability. A good deal was made of the fact that an Allied tank in the Gulf War had survived more than one hundred rounds of RPG (rocket propelled grenade) fire; rather less was made of the fact that an RPG is not really an effective tool against heavy armour. A time might well come – and arms designers are certainly working on it – when a lightweight missile will be developed that can be relied on to hit and destroy any AFV (armoured fighting vehicle) at a range at which the operator is virtually impossible to identify, let alone engage. We might assume that such a development would spell the end of the tank, but that would be premature. No sooner would such a weapon appear on the battlefield than countermeasures would be invented to nullify it – perhaps an electronic weapon that would confuse the missile.

Even if that were not the case, there would be a pressure to retain the tank. Conservative elements in military thinking would certainly be part of the problem, but political considerations might well be more significant. Politicians think they understand what a tank is and what it does, which cannot be said for a great many other military assets. There would be a degree of concern among members of the public who have an interest in war and the army. Part of the issue would therefore be a type of romantic perception, not different in principle from the resistance to abolishing cavalry regiments. There would also be a degree of self-interest. Numerous politicians represent constituencies whose economies rely, to a lesser or greater degree, on the salaries of skilled workers involved in the production of armoured vehicles. Those MPs would be most unlikely to countenance a change in military procurement policies which might bring about a rise in unemployment in their constituencies, and whether the policies were in the wider national interest would be neither here nor there.

This sort of process can be seen in other areas of military (and indeed non-military) spending, but it not just an economic matter. To many people it would simply be unthinkable that a modern army should have no tanks. Interestingly, scrapping 'big guns' due to the increasing redundancy of traditional tube artillery would probably not draw as much attention, though that might well be offset by the further adoption of missile systems for the Royal Artillery and largely pass unnoticed. These factors do not apply solely to the army. It seems increasingly likely that the development of UAVs (unmanned aerial vehicles), or drones, may soon eclipse our traditional understanding of fighter aircraft. There are several reasons why this might be the case. Drones are very much cheaper to build and the 'pilots' are very much cheaper to train – moreover they are unlikely to be killed if the drone that they are controlling is damaged or destroyed. It is, apparently, already the case that the United States trains more drone operators than pilots. We may be very close to a point where drones are so much cheaper and so much more effective than 'conventional' strike aircraft that the latter might become obsolete very rapidly indeed. That would not be the first time such a thing has occurred. The invention of steam-driven warships rendered almost every sail-powered ship in every navy virtually useless in a matter of a decade or so. That, at least, was a development that was easy for the public and the political class to accept. The end of the fighter aircraft might prove much more difficult for a reason that is strongly embedded in our cultural view of the military: a certain fondness among the public for the dashing and gallant fighter pilot.

Britain has had a long love affair with what we might call the concept of the 'artisan soldier'. Ever since the great longbow battles of the Middle Ages, the British have inclined toward the idea

Northrop-Grumman's RQ-4 Global Hawk unmanned aerial vehicle (UAV); an expensive eye in the sky, so expensive that the costs are supposed to be shared with the US by Japan, South Korea and Australia through guaranteed orders. The Chinese will be paying for their equivalent on their own.

of a small, but extremely proficient military. That concept may not have been valid, but we like it nonetheless; it has a certain romantic appeal. It has changed over the years to accommodate developments in battlefield technology or practice, but it remains broadly the same whether we are thinking about archers at Agincourt, the infantry squares and the charge of the Scots Greys at Waterloo, the 'Thin Red Line' in the Crimea, Rorke's Drift, the 'mad minute' marksmanship which made such an impression on German troops in the first weeks of the First World War or the dashing young pilots of 1940. Art and literature has a good deal to answer for, whether it is William Shakespeare's *Henry V* or Guy Hamilton's *Battle of Britain*. We might make the same observation from the opposite direction, of course. The work of a small number of First World War poets or the film *Oh What a Lovely War* have given us a clear and popular understanding of the conflict, but not one which bears more than a passing resemblance to the general course of the war.

This is not limited to one country; no doubt similar influences can be observed from the literature and films of other nations, and equally there can be no doubt that a book or a play or a painting or a film can add to our understanding of the experience of conflict. This can be the case even where the story has historical flaws. *Saving Private Ryan* is not 'based on a true story' so much as a story which has drawn on elements of several incidents, but it does present a fairly valid and comprehensible picture of war in Normandy in 1944. On the other hand, the American war hero and film star Audie Murphy went so far as to refuse to depict events which had actually occurred and which he had witnessed because he felt that they would not be credible to the public at large.

For all the reasons touched upon in this chapter – and for many others too – our perception of war at every level and in every aspect is prone to misapprehension, dishonesty, wishful thinking and cultural direction, which makes it difficult to see the nature of the problem. It does not help that political ideology is applied – deliberately or otherwise – to persuade us toward a particular viewpoint. In *The Communist Manifesto*, Marx and Engels asserted that the entire history of mankind was the history of class conflict. It certainly was not true in Marx's time.

In 1939–40 there were a number of political writers and activists in Britain, France, the United States and elsewhere who denounced the war with Germany and Italy as a tool of capitalism, which was being applied to make the rich richer and the poor ever more subject to the will of a cabal of international capitalists.

From a capitalist perspective, the reality of war (as opposed to the commercial opportunities arising from continual preparation for it) is a disaster. It may bring astronomical profits in the short term, but a war has to be paid for afterwards and that almost inevitably means higher taxes for everyone. The richest people can generally arrange their affairs to minimise their tax burden, but most people cannot. If the people have to bear a higher tax burden they – obviously – have less money to spend on the goods and services which capitalists provide in order to generate wealth. A relatively short period of wartime boom is more often than not offset by a lengthy period of post-war austerity and a slump that can quite easily degenerate into a depression, which may heap yet more immediate personal pain on individuals at the bottom of the economic ladder, but can also bring about the ruin of even the largest commercial enterprises.

Even so, war can be an economic catalyst. It was the Second World War, not Roosevelt's New Deal, which rescued the economy of the United States and made America the dominant force in global production and politics; but the recovery was built on the spending of government, not the investment of capitalists. That has not prevented a great deal of talk and literature about how free enterprise saved the world from totalitarianism.

Both the Right and the Left of the political spectrum are happy to claim that war is the tool of the other side. In fact, political groups of all denominations are willing to take to war as a means of achieving an objective. They may not have a choice; war may be forced upon them by an aggressive neighbour and the desire to survive, but all nations, all societies, all faiths and all political ideologies are quite capable of waging war if they feel it is in their interests to do so, and all governments – democratic, oligarchic, autocratic and theocratic – are prepared, if necessary, to persuade, cajole or coerce their constituents into serving in combat. (Buddhism, you cry? The civil war in Sri Lanka was fought between predominantly Buddhist Sinhalese forces and Hindu/Christian Tamils.)

'History repeats itself.' It really does not, but we can see similarities in different situations which might lead us to think that it does. More often than not, the similarities are imposed by our assumptions. Many British (and other) historians and military writers have drawn parallels between the reliance on marksmanship that distinguished the longbow archer of the Middle Ages and the highly trained riflemen of the army of 1914 from their counterparts in other armies. The similarity is clear; a small number of skilful men inflicting defeat on a much larger force. The fact that neither example actually led to victory in war (as opposed to victory in battle) is conveniently set aside in favour of a romanticised view of the value of marksmanship. The distinctions do not stop there. The infantry soldier of 1914 was trained to employ fire and manoeuvre to kill the enemy at a distance. The role of the longbow archer was to disrupt and weaken the enemy so that he could be finished off at close quarters by the infantry. The regular soldiers of 1914 were, in the main, career professionals; their longbow forebears were, in the main, men who joined the service for a short period of adventure. Similarities in given situations – even when they are real rather than imagined – are inevitably outweighed by the differences.

If there is one popular military maxim which actually does carry its own weight, it has to be 'no plan survives contact with the enemy', though there are still caveats. All in all, plans should not have to survive in perfect entirety. The purpose of a plan of any kind, whether it is for an attack, a defence, a withdrawal or to frustrate the intentions of the enemy, is to achieve an objective. As the dispositions and intentions of the enemy become more clearly identifiable, and as those dispositions and intentions change in reaction to developments in the field, alterations to 'the plan' will almost inevitably be required if the objective is to be secured and success – if it is forthcoming at all – to be exploited to the full.

This is likely to be the case even if 'the plan' is successful; in fact, it may have to be adjusted more radically and more quickly if the plan is more effective than originally envisaged because more and better opportunities are likely to arise if the enemy is comprehensively defeated and his forces thrown into a state of confusion. The collapse of the German army in France in the summer of 1944 gave the Allied armies an opportunity to make incredibly rapid advances; so much so the Allies had more opportunities than they could possibly exploit. There were simply not enough troops to take advantage of all of them, and even if the men and material had been available, the means of delivering adequate quantities of logistical support was not.

PART 3

CONCEPTS, INSTITUTIONS AND PRACTICAL EXAMPLES

8. Command and Staff

If there is to be a viable military force at all, let alone one that functions effectively, there must be a command structure and that structure must have a specialist staff of people whose function is to make the decision process viable. The commander must make those decisions and take responsibility for them – in fact, he will often have responsibility dumped at his feet for actions that were none of his doing or which were imposed on him against his better judgement. The opposite holds true as well. From time to time, commanders have been given credit for success which occurred despite their actions rather than because of them.

The precise process of decision-making depends very much on the traditions, culture, experience and structures of the army, but specific operational decisions also depend on the nature of the individual commander. Some have relied more heavily on advice and intelligence than others. Some have relied on methodology and ideology, some on confidence in a particular approach or a particular weapon system. Some have even relied primarily on the incompetence of the enemy and some have just battered on blindly and hoped for the best. All of these aspects have brought success and all have brought failure. Many books – and countless lectures – have been produced describing what are often called 'principles of war'. The term is not especially good, since there is very little that we can honestly say is universal, and therefore they are not so much principles as guidelines, but there are processes which, when applied rationally, help to give commanders a view of their situation, which will hopefully assist them to make 'good' or 'sound' decisions. We should shy clear of the concept of the 'right' decision.

A battlefield decision or policy – or a strategic one for that matter – is only 'right' once it has proven successful at a given time and place. That does not mean that it was the only option or even the best one. It may only have been 'successful' in the sense that it did not bring about a catastrophic defeat, though in some circumstances that in itself may have been a remarkable achievement. The commander who manages the most difficult of operations – to make a secure withdrawal against heavy odds to a location where his army is safe from destruction but continues to pose a serious threat to the enemy – may have done a remarkable thing and demonstrated outstanding professional skill but be regarded as having been defeated since he chose to give up ground to the enemy.

Wars unfold in a manner that cannot be accurately forecasted. However much information is gathered about friendly and opposition assets or about the terrain, the climate and every other factor which might impinge on the course of the campaign or operation, there will always be a host of issues which affect the outcome. Some of these will have been considered, some may have been recognised and understood, it may have proven impossible to take any action in regard to them or the actions taken might turn out to have been less effective than hoped, irrelevant, inappropriate or even counterproductive. Many issues will not be recognised at all until they start to have an impact on the conduct of the battle and many more may not be recognised until after the operation has been concluded, if at all.

The business of making a plan is, therefore, extremely difficult, and decisions have to be made with a limited understanding of every aspect of the process. Various techniques or practices have developed to help commanders with the decision-making process. They can only help; there is no system for prescribing success. There are any number of collections of 'maxims' to guide commanders, and most armies, if not all, have had systems which have given better results – as a rule – than no system at all. Each of these systems has had its own terminology, but naturally they mostly cover much the same ground. There may be significant differences in the approach, of course. In the British tradition, topics such as camouflage, concealment and deception have been addressed as functions of tactical thinking, whereas the Soviet tradition has presented them as aspects of a military subject in its own right – *maskirovka* – and a practice to be applied in all situations for its own sake, not simply as an adjunct to a given manoeuvre or operation.

Regardless of the ethos of his training, every commander must take on board the general condition of his position and that of the enemy, what we might call the military environment. Without an understanding of where he is and what he is trying to achieve and the means at his disposal any plan is very likely to fail – it may not even be possible to mount the operation in the first place. Every sound plan must have a basis in what is possible and what is probable, whilst making allowances for the unidentifiable and plain common-or-garden bad luck. In some cases the environment will be familiar to the commander, particularly if he is fighting on his home turf, but very often – even if he is – the requirement to plan and execute operations will still remain. The opposition may have made an unexpected invasion, or he may have been appointed to a command or given a task that he had not envisaged. In each case, the commander must take action to carry out the instructions of his government and to protect his force.

Whether the operations are to be defensive or offensive in nature, he must formulate a plan to achieve success. On occasion, his plan might be no more than to avoid contact with a superior enemy and thus avoid defeat, though, even in the short term, failure to develop a plan for victory will very often end in defeat anyway.

In the course of the planning stage, the commander must understand as best he can his environment, the existing commitments of, and threats to, his own assets and those of the enemy.

He needs to identify the objectives of the enemy as well as his own and to make a plan that will further his own ends whilst preventing the enemy from gaining his own objectives. He must, therefore, ensure that the men who will be trusted with executing his plan really understand what is required of them and that they are aware of the wider picture but not distracted from their own responsibilities. Those men must be able to make valid and rapid decisions during the operation, but they must also be capable of judging what they will require to carry out their own task and ensure that all the necessary training and rehearsal has been carried out to the required standards.

Once the plan has been decided and the troops have been briefed and the operation has been launched, the commander has to be constantly vigilant that his subordinates are pursuing their objectives in the best way, that they are not digressing from the main task and that they are not being either too cautious or too bold, and that they are not being profligate with manpower or materiel. Throughout the operational period the commander also has to be extremely vigilant; watching for new initiatives from the enemy, changes in the weather, to react against unexpected threats and to exploit unexpected opportunities. Meanwhile – naturally enough – his opponent will (or at least should) be doing everything he can to prevent the success of the very operation that our commander has undertaken. One could make similar observations about the world of commerce, of course; developing new products, acquiring new markets and undermining the opposition's activities are natural parts of the business environment. The great difference is that in business (unless you are in the mafia), neither party is permitted, never mind actively encouraged, to kill the operatives or blow up the installations of the other.

However clever and innovative a plan might be, it will not be possible to put it into action if there is no means of relating resources to tasks and delivering instructions to subordinates.

The commander must have a staff. Staff officers take a long time to train because the process of finding the right resources, getting them to the right location and ensuring that troops with the requisite abilities are on hand to use the equipment properly is immensely complicated. A larger army is likely to have more complex demands, though this is not always the case. There are many examples of large forces with basic weaponry and unsophisticated needs having to face up to a smaller force with more sophisticated equipment and practices; most colonial wars have fallen into that category. If the 'simpler' force is to achieve success it may actually have to have better staff work in the sense that better timing and/or a more painstaking approach might be critical since there are more limited resources. The more 'complicated' force will generally have a much wider range of demands which must be met if the whole system is to work properly. Competent staff work is the real foundation of success. However brilliant the commander might be, however innovative his approach, the chance of success is very small if the right troops and material are not concentrated at the right place and the right time. There is little value to having superb tanks if they cannot be refuelled on demand and the best rifleman in the world is useless as soon as he has fired his last bullet. The requirements are much more demanding than simply providing food, fuel and ammunition. Broadly speaking, the larger the army, the more demand there will be for a wider and wider list of stores and specialists to maintain the efficiency of the force.

As armies have become more dependent on technology, the demands on staffs have increased exponentially. Once armies became mechanised a whole new range of assets were required. It was not just a matter of providing drivers and fuel, but of providing mechanics, engineers and workshops which could be kept close enough to the front to offer effective support. These support elements in turn require a steady stream of specialist supplies; they need constant re-training and re-equipping to keep them up to date with new vehicles or improvements to old ones. They need to be fed and to have medical services, and more often than not the whole structure has to be able to move to new locations rapidly and with very short notice. As if that was not a complicated demand in itself, the whole business has to be conducted under threat of enemy action and frequently in an environment that is at least marginally different from – and sometimes quite alien to – that to which the troops are accustomed or for which the army has been prepared.

Naturally, there is nothing particularly 'modern' about the principle. Ever since the days when armies started to consist of anything more than a handful of men wandering the countryside in search of better hunting grounds, wise leaders have made preparations for conflict, and the process has been complicated for a very long time. When Henry V took his army to France in 1415 he ordered a quarter of a million arrows. The arrows obviously had to be carried with the army, so there was a need for barrels of a given size to carry them in. The barrels required wagons, the wagons required horses and the whole business therefore had to have dedicated operatives – drivers, ostlers, farriers, blacksmiths, harness makers – and of course tools, materials and foodstuffs for men and beasts alike which in turn called for more wagons, animals and men. Even for a weapon system as simple as bows and arrows, Henry's staff had to acquire and organise a substantial train of assets just to ensure that the arrows would be available at the point of need.

There is an adage used frequently by journalists reporting from war zones: 'Amateurs talk tactics and professionals talk logistics'. It is a neat phrase and has a certain appeal, but in reality the two are dependent on one another. The tactical (or strategic) planning process does depend on the capacity to deliver stores at a given time and place and in a given quantity, but the logistical demand is also shaped by the strategic and tactical demands of the theatre. It is extremely difficult even for a modestly sized military establishment which is strictly dedicated to the defence of the national border, and obviously that much more of a challenge to those countries which require – or aspire to – a military establishment that can project power throughout a region, and an even greater challenge to those which want to have a worldwide role.

Even for the country which focuses exclusively on what we might call 'homeland defence', the nature of the potential threats for which they must be prepared is not always clear. There may be more than one neighbouring power with very different approaches to war, very different armament and methodology. The border itself may run through a variety of environments, each of which presents its own challenges. An invasion through mountain passes is a very different beast to one across a plain or which requires a response to an amphibious landing. The threat may not even come directly from a neighbouring power but from internal dissent and insurrection, which would demand quite different approaches militarily and very probably politically and socially as well. Intensive bombing against a border incursion might be acceptable to the population as a whole, but the utter destruction of a town or city inside the national border (even in operations against a foreign aggressor) might not be a valid option.

9. Concentration

We may comfortably discard the easy application of the 'God is on the side of the big battalions' concept when it is applied in a loose and casual way, but – like most clichés – there is a degree of truth to it. As Lenin put it, 'quantity has a quality all of its own'. Mere numbers alone in the simplest sense are not a useful guide to success; they can even be detrimental to the effectiveness of a force. If there are more troops present than can be usefully applied, they may be more of a liability than an asset. They may not be making a contribution to the operations, but they still have to be fed, clothed and paid. They have to be protected from attack and they have to be kept occupied so that they do not become disaffected – 'what are we doing here' – and even possibly mutinous. They can become an unnecessary and unwelcome drain on the resources of the rest of the army or on the state.

Local superiority is, however, a very different matter. It is not necessarily significant that the enemy has much greater numbers of troops and tanks and planes and ships; what matters – in the immediate future at least – is where those assets are and whether they can be applied in time to have an impact on the outcome of the current operation. What is important is how many troops, and with what degree of competence, can be put into the field at crucial locations. The ability to outnumber the enemy at a specific spot does not ensure victory any more than any other factor, but it is certainly an advantage. The question is how it can be achieved. Even if a given commander has a very much larger force, it is not inevitable that he can bring about a concentration that will give him local superiority; that will depend on several factors.

The commander may have a range of obligations that he cannot ignore which may be military, but might be political, religious or social. A location of great spiritual or cultural importance to the country as a whole may have no military value whatsoever. It may be impossible to abandon it to the enemy for fear of damage to national morale or because it would undermine credibility in the ability of the government. The defence of Malaya in 1941–42 was conducted with a view to protecting the massive Singapore naval base which had been built at gigantic expense to provide facilities for a British fleet which would protect the shipping lanes from Australia and New Zealand to India, the east coast of Africa and the Suez Canal. Due to the demands of the war in Europe and North Africa there really was no fleet to speak of. The only two major assets – HMS *Repulse* and HMS *Prince of Wales* – were sunk two days after the opening of the campaign and the naval base had no real value thereafter, but it still had to be defended because its loss would have been a blow to the government in London.

The commander on the ground – General Percival – was probably never capable of conducting a successful campaign against the Japanese invasion force, but the obsession with protecting the naval base certainly tied his hands by preventing him from concentrating his forces. Even before the invasion commenced, Percival had already had a similar problem forced upon him by the policies adopted by the British military authorities. It had been decided that any defence of Malaya would depend heavily on air power and accordingly a great deal of effort and money had been spent on constructing major air bases and dispersal airstrips all over the country. The principle may have been right or wrong but, since the British government failed to make an adequate commitment of modern aircraft, the bases were fairly useless – but had to be defended – which was clearly a task for the army.

Accordingly, a great deal of Percival's force was spread out in small packets over considerable distances to protect airfields with no aircraft. It quickly became apparent that there would be no immediate reinforcement of fighters and bombers that might slow the Japanese advance, so why did Percival not concentrate his troops along the roads that the Japanese would have to seize in order to march on Kuala Lumpur and Singapore? He did make some effort to do so, but he was hampered by a number of obstacles, not least of which was poor basic staff work, but there was also a shortage of transport vehicles and a shortage of fuel – or, more accurately, a shortage of these things in the right locations. It might seem a safe assumption that the trucks and the fuel might have been brought to the right places at the right time with a modicum of effort, but Percival was also hampered by Japanese fighters and bombers which – to add insult to injury

– were operating from the very airfields to which army units had been deployed and which had been abandoned due to the lack of aircraft and the rapid advance of the Japanese ground forces.

Failure to concentrate his forces was not Percival's only problem, but concentration itself is not a silver bullet solution to challenges of command. It is only an effective contribution to victory if the troops are concentrated in the right place, in an effective manner, in appropriate combinations of different arms of service, with the necessary stores and equipment, and at the right time. A major concentration of resources in one area may result in an inadequate distribution of troops elsewhere and give the enemy opportunities to strike significant blows with a relatively small number of men.

Concentration, like most concepts in war, is applicable at every level of command, not just to the actions of armies over large territories. It can be just as significant to the junior commander who must decide how he approaches a relatively small objective such as a line of houses held by the enemy. Should he concentrate his firepower on one or two houses, force the enemy from that position and then occupy it so that he can then direct his men to fire down the line of the enemy's remaining positions? If so, should he try to take a house or two at the end of the line or in the middle? Even at the level of a single squad of eight or ten soldiers, the question is not simply whether to concentrate effort, but where and to what purpose? An engagement of any size is not an end in itself, save in those rare instances where the engagement might result in the complete and utter destruction of the enemy, or at least in his ability to continue fighting; the commander must consider what is to happen next. If he concentrates his resources to deliver a more powerful attack will he – if successful – make his force more vulnerable to the reaction of the enemy? Might his forward movement put him at risk of isolation? Will it put undue pressure on his lines of communication? Will the operation yield results that are in tune with the wider strategic goals?

V for victory from a British Indian Army soldier as he arrives at Singapore – and according to the original caption, 'his "V" is backed by a million Indian troops and the rest of the Empire as well'. He probably disembarked just in time to surrender.

10. Economy of Force

The term 'economy of force' appears in a great many treatises on war theory and seems to be rather obvious. It might be seen as a recommendation for the commander to use the smallest possible force to achieve his aims, but is really a rather more sophisticated concept. In war – as in commercial economy – the aim is not necessarily to apply the minimum level of resources, but to apply resources as efficiently as possible. A smaller force may be quite capable of carrying out a mission effectively, but might leave itself vulnerable to heavier losses than might otherwise have been the case, or find itself open to an immediate counter-attack having exhausted its capacity to sustain further action due to exhaustion of the troops or of ammunition, fuel or other material.

Economy of force may, therefore, require a much heavier commitment than is strictly required for the task in hand. There is little value in achieving an objective only to be forced to abandon it – or for the force to risk being severely damaged or even perhaps destroyed – by the reaction of the enemy. The mere presence of a very strong force may actually remove the necessity to fight at all. If the enemy is aware that an overpoweringly strong force is approaching he may feel that the best option is to abandon his position or his current intentions. Clearly, that would be a very effective use of force in that it did not result in any losses whatsoever, but did bring about the acquisition of the objective.

If success without engagement is the most effective application of force, we must give some thought as to why the concentration of a massive force at a single crucial location is not the solution to every tactical problem. The first barrier is assembling the force. It is very rare for a commander to have so much force at his disposal that he can commit a massive force in one location without denuding other parts of his operational area. If that disparity is achieved at all, it is very often the case that the conflict, or at least a specific campaign or series of operations in a given area, has actually already been won to all intents and purposes. If the enemy is so weak that he can be safely contained and countered in every part of the theatre in addition to the one where he is massively outnumbered, his prospects must be – at best – pretty poor. More generally, commanders must have a grasp of economy of force in order to conduct successful operations at the crucial locations because they cannot acquire enough men and material to mount a continuous offensive everywhere at the same time.

Economy of force is not, therefore, a matter of sheer numbers, but of the best application of what is available under the circumstances and – most importantly – the application of the right sort of resources in the right proportion. Most students of war are – or at least certainly should be – aware of the concept of 'combined arms', meaning that success on the battlefield requires the efforts of various forms of weaponry in support of one another. The proportions of those weapons required for force efficiency is not constant. Armoured vehicles are more vulnerable in close country or urban conditions where anti-tank weapons can be concealed until the vehicle is in effective range, and a predominantly infantry force is more vulnerable in flat, open country where they can be engaged at long ranges by artillery, aircraft and automatic weapons. Almost every form of terrain will, however, demand the application of all the systems available to the force commander and he will seldom have complete discretion over the force entrusted to him. Clearly, this is a fairly major challenge in itself. The force that roamed freely over the plain with its tanks and armoured personnel carriers making rapid advances may find itself in a very unhappy situation when it reaches a town or city that cannot be bypassed and must be cleared of enemy resistance.

It is a simple point to make, but a hard one to deal with. A large body of troops develops its own operational ethos. The subordinate commanders become accustomed to one another's ways and they learn which of the different units in the force can be relied on in a tight corner – and just as importantly which ones cannot. Changing the composition of the force is not as straightforward as removing one unit of tanks and replacing it with one unit of infantry or artillery. The decision whether or in what circumstances to add, subtract or replace units may not, therefore, be quite as easy as deploying more infantry or armour or artillery units, which would be more appropriate to the changing environment of the battle. There may be a cost in efficiency and cohesion and possibly to the morale of the force generally when a new unit is introduced and an existing one removed.

There is another and rather different aspect to economy of force. Committing a small force to a given area or operation may not be adequate to achieve an outcome in the sense of taking an objective or holding the enemy at bay, but still be a useful deployment. Assuming that the enemy is unwilling or unable to ignore that portion of the theatre or the battlefield, a commander may be able to draw enemy resources to that location which could be put to better use elsewhere. As long as the force committed to that location is modest in comparison to the resources that the enemy has to deploy to counter them, that would certainly be an economic use of force, but it is also possible for a large force to be an economic proposition when deployed against a relatively modest opposition.

If the resources of the enemy are already stretched thin – or just even a specific type of resource – there may be an advantage to be gained in another part of the battlefield. If, for example, the enemy has few tanks, forcing him to divert even a small proportion of them to a less critical location may give the commander's own armoured units a significant advantage that can be exploited at a crucial place or time. This principle can be identified in aspects of naval warfare in the Second World War.

The primary mission of German assets such as the *Scharnhorst* or the *Bismarck* was to destroy and disrupt Allied merchant shipping. These were immensely powerful ships, far more heavily armed and far more expensive to maintain than the destroyers that would have been perfectly adequate for dealing with merchant ships, but since they were so heavily armoured and gunned, the German battleships were more than a match for convoy escort vessels and therefore drew far greater numbers of Allied battleships and cruisers away from other tasks. The relatively small number of German battleships and battlecruisers would not have been at all effective if committed to seeking a major battle with the Allied fleets; they were very fine ships, but would have been completely outnumbered, but as commerce raiders – often operating singly – they not only sank huge quantities of vital ships and the supplies on board, they forced the Allies to go to great efforts to hunt them down.

11. Defence or Offence?

At its most basic, the purpose of war is either to conquer or to defend, though it may be hard to divorce one from the other. Simply defeating the enemy on your own border might conceivably be sufficient if you are on the defensive, but a successful border action is seldom if ever adequate for the attacker.

Even so, as a defender inflicting a defeat that repulses the enemy and preventing him from advancing into your territory will not generally do the trick. In most cases, the enemy will already have acquired a portion of your territory in the initial phase of the battle and political reality will demand that that portion is recovered as quickly as practical. Even a major success on the battlefield may not be enough to send the enemy back into his own country. Taking the First World War as an example, once the Germans had occupied most of Belgium and a sizeable portion of France the Allies were faced with the prospect of ejecting them. The military conditions of the period made this an extremely difficult challenge. The effectiveness of machine guns, artillery and fortifications in the form of trenches and wire entanglements gave a very real advantage to the defence over the offence. This led to a massive and costly war of attrition which strained the physical, emotional and financial resources of all of the participants for four years. The Allies managed to secure several major victories on the battlefield, but the gains made were trivial in comparison to the losses incurred.

In short, the Allies were obliged to conduct a defensive strategy with offensive tactics and the situation was made worse by what were seen as political imperatives. The French government were determined that no territory – however small – should be given up to the enemy. The political/ideological rationale may or may not have been valid, but it was militarily senseless. French commanders were prevented from making any kind of sense of the Forward Edge of the Battle Area – or FEBA. This is a term that students of war will encounter in many books, papers and articles and it is worth taking a moment to explore it. It means – rather obviously – the point of contact with the enemy; the 'sharp end', on a smaller scale, the *Schwerpunkt*. Success in battle depends on attaining supremacy over the enemy, but that is not simply a matter of numbers or better training, equipment or motivation. Geography, or perhaps more specifically, topography, is a crucial factor in gaining and retaining significant advantage in FEBA, and thereby having the upper hand in deciding the location of operations.

Once the Germans had acquired territory in 1914 the onus was on the Allies to drive them out, but the Germans could afford to set up a strong defence on French and Belgian soil and let the Allies batter against it. They could afford to abandon weaker positions and focus on stronger ones, typically having the forward edge of their deployment on ridge lines – even very slight ones that scarcely show on a map – which overlooked Allied forces. Consequently, they could observe the movement and deployment of French and British troops and direct artillery fire with considerable accuracy. Just as importantly, they could move reserve units with relative impunity since the bulk of their forces were concealed from view. As long as the Germans held French or Belgian territory they could claim to be 'winning', which had an understandable effect on morale both at home and on the battlefield. At the same time, the inability of the French and the British to make major inroads against those positions meant that the Allies were not winning, and if you are not winning you are probably losing.

Obtaining a significant advantage in FEBA essentially handed the initiative to the Germans. The term 'initiative' is often construed as the ability to force the pace of the battle and to make the enemy respond to your actions, rather than the other way around. In a crude sense, this is certainly the case, but it is not as simple as deciding where or when to mount attacks. The Germans could not prevent the British and the French from mounting operations, nor could they force them to choose particular sites to attack, but their retention of a highly defensible line on French and Belgian territory meant that the Western Allies had to attack if they were to make inroads on the German positions. The tactical initiative, in a rather weak sense, lay with the French and the British insofar as they would be mounting the offensives, but the strategic initiative lay with the Germans in the rather stronger sense that they could prepare for attacks that they knew had to come somewhere. So long as they could maintain strong reserves close enough to the front to intervene quickly, but far enough away to be out of immediate danger from shelling or being drawn into the battle prematurely, the Germans would enjoy a considerable advantage.

That the policy was not, in the end, successful has more to do with the overwhelming demands on German manpower and industry and of course the entry of the United States into the war in 1917, but it did meet the needs of the German political strategy quite well for four years. This is more of an achievement than is generally recognised in Western Europe and the United States; it is easy to forget that Germany fought and won a war on a gigantic scale against Russia while holding off the Western Allies. Of course, it is true that the Russian Revolution brought about the end of the war in the east, but the success of the revolution was, to a great extent, a consequence of a failed war. This is not to say that the revolution would not have taken place at some point, but the war certainly hastened the process and provided the political and economic environment that brought it about at a particular juncture, which, as it happened, favoured the Germans since, with Russia out of the war, they were able to transfer assets to the Western Front.

12. Focus; the Maintenance of an Aim

Strategic and tactical aims can only be achieved by identifying a course of action, assembling resources and constructing viable schedules for the acquisition (or denial) of objectives. Once the aims have been identified and the plan put into action, it is crucial that the commander keeps his eye on the ball, that he is not distracted from his objectives and ensures that his plan is completed. That all seems fairly obvious and indeed the planning process should always be rooted in a careful and realistic appreciation of what forces are available, time constraints, the significance of the objectives and a rigorous examination of the potential benefits and risks of the operation. All of the complicated and demanding analysis that goes into mounting an operation will come to naught if the commander and his subordinates lose sight of the goal. Equally, the commander must always be conscious of the ever-changing nature of the military environment. An unexpected change in the weather or an unexpected opportunity and – above all – the response of the enemy to his actions or a new initiative mounted in another part of the battle area may force him to alter or even abandon a plan which seemed to offer significant gains. This may be the case even in instances where the operation as planned has been developing in a very satisfactory manner. It should be fairly obvious that an advance which is making good progress and has great promise may have to be adjusted in the sense that additional resources may have to be applied or the attacking force halted or even withdrawn if the enemy send strong reinforcements to the threatened location. Altering the objectives may be a consequence of identifying a strong risk of failure in the initial offensive, but it can just as easily be required – or at least desirable – simply because the enemy has made what appears to be a successful realignment of his forces. Every commander is limited by the strength and nature of his force, but the same constraints apply to the opposition. If the enemy has reinforced his deployments in one area, it will generally have been at the cost of another location which may – in its weakened condition – constitute a better target. Making a judgement about altering the thrust or nature of an operation is an extremely hazardous business, but failing to seize an opportunity may be disastrous. Maintaining an aim simply for the sake of doing so can be a failing rather than a virtue. Commanders tend to be loath to give up a plan and can easily delude themselves into the belief that 'one more push' will bring the results they were hoping for. One of the hallmarks of the superior commander is the ability to consistently make sound judgements about the current viability of his plans, knowing when it is time to press on vigorously and when it is time to accept that he should cut his losses. That is an immensely difficult decision to make and one that is affected by considerations beyond immediate tactical advantage. Failing to press on with a particular initiative may, as we have seen elsewhere, have a damaging effect on the morale of the troops, the confidence of subordinates and of perceptions 'back home'. Any of these may be more significant than failure to attain a particular outcome, particularly so if the initial objective was not of paramount importance. The commander will – hopefully – have access to sources which can help him make the decision, most importantly intelligence and analysis of the condition of the enemy. However skilled and conscientious his intelligence staff may be, they will never be able to provide him with an absolutely perfect picture of the enemy's resources, intentions or options. In the end, he will have to be guided by his own experience, his general 'feel' for the course of the fight and, perhaps most importantly, his understanding of the ability of his own forces to respond to a change in plan.

13. Combined Arms

'Combined arms' is a term that has gained extensive currency among journalists and others in their descriptions of conflict and is often seen as a modern concept, something that applies to the age of mechanised warfare. The tanks, aircraft and artillery combine in differing, but complementary, roles with the infantry – a truly twentieth-century phenomenon. Nothing could be further from the truth. It has been the rule rather than the exception since the Bronze Age and its roots lie in the development of the warrior class. Although the warrior was expected to justify his existence through his greater commitment to training, his acquisition of superior arms and armour and his readiness to fight, the general assumption was that in times of greater peril the rest of the community – or at least the males – would turn out for battle as required. In some cases, the warrior class would take the field in the same tactical organisation along with everyone else and provide leadership and inspiration to the group. His superior armament would benefit him and, indirectly, those around him. He could serve as a sort of tiny strongpoint amid the ranks and files of the formation.

This was not necessarily the best use of the warrior. It could be more effective to concentrate a proportion of the warrior class into a single body which would have greater combat power due to better armament and training. The warrior unit might then be deployed as a spearhead to break the enemy formation or as a reserve to be committed at a crucial juncture once battle had been joined.

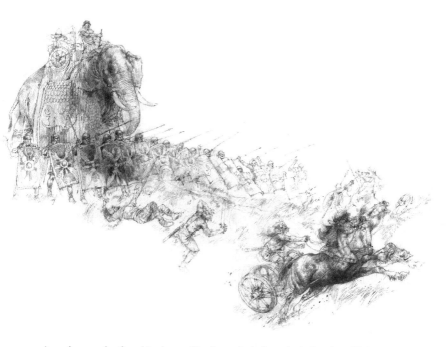

An early example of combined arms; Claudius took elephants in the invasion of Britain of AD 43 to awe the population and terrify the chariot horses. (Victor Ambrus)

The combined arms concept had become a normal aspect of combat by the time humans started to record their wars. Ancient artworks clearly show distinctions between differing troop types – bodies of archers or slingers, bodies of close-combat troops with spears or axes, bodies of cavalry or chariots.

Every type of weapon has different strengths and weaknesses so the ability to commit units of spearmen or archers as discrete elements had a very real value in battle. A body of spearmen in a tight formation – if they held their nerve – would be stronger than a body of cavalry which had trained for a charge to contact, whereas a body of archers could inflict steady losses on the spearmen so long as they could avoid becoming embroiled in a close-quarters fight.

Even the rare exception of a society which produces only one sort of warrior, such as the cavalry armies of Genghis Khan, still conforms to the combined arms model, it's just that the warrior has a wider range of skills and weapons. Depending on the situation, the steppe warrior could strike with the lance or bow, but his horse gave him the speed to evade any serious threat. Naturally, he could only use one weapon at a time, so we can identify a practice of having one or more units weaken the enemy with their arrows while another stood ready to deliver a charge to contact. The combined arms principle still applied, but all the elements of the force could be deployed to any of the battlefield roles that the army employed.

The primacy of one or other form of soldier in different periods of history is widely acknowledged but is something of a myth. It is, for example, an article of faith that the warrior class of Europe in the Middle Ages, the armoured cavalryman, was all that mattered and that the infantry were an afterthought, mere cannon fodder. That begs the question as to why the overwhelming majority of soldiers in every medieval army fought on foot; indeed, it became a commonplace for the 'knights' to dismount for battle. If nothing else, since war is such an expensive business, why would kings expend such large sums to pay and feed thousands of infantry if they were not a crucial part of the structure?

We can see the combined arms principle in a single formation when we look at the wars of the sixteenth and seventeenth centuries. A single regiment would consist of pikes and muskets. Despite many fine, dramatic Victorian depictions, the musketeers did not form up with the pikes reaching over their heads; the pikes were arrayed in a central body with the musketeers to the left and right. The role of the pikes was to provide secure flanks for the musketeers. In due course, the pike disappeared as musketeers started to use bayonets and thus acquired a double function as both close-quarter and missile troops. The two roles were combined in each soldier – no different in principle from that of the steppe warrior with his lance and bow.

The ability to combine systems does not mean that there are never situations where a commander may choose to deploy only one form of force; there are circumstances where the cavalry or the tanks might operate independently for the sake of speed or be rendered ineffective because of the terrain. The whole point of the combined arms concept is that it provides the commander with options.

14. Flexibility and Versatility

Being flexible is not the same as being mobile. Clearly, there must be some degree of overlap if only because an immobile force cannot be an altogether flexible one, but it is not difficult to conflate the two terms. An airborne or amphibious unit can be landed in any number of locations and is therefore self-evidently mobile, but the structure of such a force actually prevents it from being as flexible as it might be. Airborne and amphibious units tend to be smaller than 'conventional' infantry battalions – often with only three platoons to a company instead of four and/or only three companies in the battalion rather than four. The individual squads and platoons may be numerically weaker and the unit will – as a rule – have fewer and/or smaller integral assets along the lines of an anti-tank platoon or an engineering section. In that sense, the 'conventional' battalion is the more flexible in that it can be more readily applied to a wider range of operations.

Field Marshal William Slim, one of the great soldiers of the twentieth century, was not keen on the concept of 'specialist' infantry forces. In this context, it is important to distinguish between 'specialist forces' and 'special forces'. The latter applies to very small bodies of men who are exceptionally highly motivated and who are trained for a range of tasks such as infiltration, sabotage and intelligence-gathering operations. We can use the term 'specialist forces' to describe formations with a more conventional combat role such as parachutists, marines and rangers. Slim's belief was that any well-trained and well-led infantry battalion can be applied to any battlefield task with a modest amount of extra training and, where necessary, equipment that is relevant to that task.

At a higher level, flexibility is achieved – or at least aimed for – by the ability to organise subordinate forces of varying composition according to the need of the situation. If a Division of six infantry and six tank units finds that its area of operations includes a wide expanse of open country divided by one or more dense forests or urban areas, the commander can allot more armour to one part of his front and more infantry to another. In the past – and perhaps still – it was common to describe the brigades of an armoured division as being 'tank heavy' or 'infantry heavy'.

There are other factors, of course. The existence of the 'specialist infantry' force can provide the commander with opportunities to seize an objective or disrupt the enemy by effecting a landing behind his lines; airpower or chemical or biological weapons can force him to abandon operations or positions, or electronic warfare assets can incapacitate his communications. All of these things, and a great many others, give the commander flexibility in both the range and nature of initiatives he can undertake. None of them will be of any great value unless the commander himself is flexible in his attitudes and versatile in his approach to battle.

15. Casualties

We are all aware that war involves death and injury; if there is fighting then there will be casualties. The term is rather loosely applied by the press, historians, politicians and, not infrequently, by the military as well. At one extreme we might consider a 'casualty' to be a person who has been killed outright and at the other we could interpret the same term as describing individual soldiers who are not immediately available for combat – soldiers who are absent through sickness, injury, death, or desertion or who have simply become temporarily detached from their units in the course of battle.

Although killing and wounding are rather obvious factors in combat, they are very seldom the determining factor in victory or defeat. There are a few salient exceptions; Thermopylae and the Battle of the Little Big Horn are two rather obvious examples. If all of the soldiers on one side are killed that is a pretty clear indication of defeat.

Heavy casualties will, generally at least, have an impact on the course of a battle and/or our perceptions of the abilities of the commander and of the competence of the force. It is rather harder than we might think to make a valid assessment of what constitutes 'heavy' or 'light' casualties in a given situation. The Battle of the Somme is widely regarded as an example of brutally heavy casualties. Any number of writers have commented on the loss of 60,000 'British' soldiers – or even in some cases 'English' soldiers – killed on the first day of the battle. This is simply incorrect. The British and Commonwealth/Empire forces did suffer about 60,000 casualties, of whom about 20,000 (estimates vary, but not greatly) died on the day or from wounds. Either figure is certainly depressingly large, but do they represent 'heavy' losses? The Commonwealth/Empire forces committed on the first day of the battle amounted to something in the region of 750,000 people.

The nature of the operation is probably best seen as an attempt to storm a gigantic fortress as opposed to a battle of manoeuvre, and such operations tend to be very bloody and result in very modest territorial gains even if they are completely successful in the sense of acquiring all of the objectives. The attacker must leave his own positions to attack a fortified emplacement, advancing over countryside consisting of fields of fire which have been carefully assessed and marked out by the defenders. Assuming even a very basic level of competence among the defenders, it should come as no surprise that the attacker's losses will be considerable.

How then, does the Somme compare with other massive battles? The Allied army at Waterloo – more than half of the soldiers were Dutch, Belgian or German – consisted of about 68,000 soldiers, of whom a little over 10,000 (1 in 7) were wounded and another 3,500 (1 in 20) were killed. Of a little less than 40,000 Confederate soldiers at Antietam around 1,500 (1 in 16) were killed and 10,000 (1 in 4) were wounded. Even if we ignore the non-divisional elements which were involved in the battle, the comparable casualty rates for the Somme were in the region of 1 in 37 killed and 1 in 19 wounded. It would be right and proper to ask whether these battles are genuinely comparable. In one sense they are not. The battle area of the Somme engagement was very much larger and, of course, the battle continued for months on end with hundreds of thousands more men wounded and killed. We might make the point that many individual units suffered exceptionally high casualties – 50 per cent and more – but the same can be said of Waterloo or Antietam. On the other hand, Waterloo effectively ended the war against Napoleon – there were more actions, but none of any real significance. The Somme caused irreparable damage to the German army and the Allies learned some of the lessons that would bring victory in 1918, but Antietam achieved virtually nothing. The 'butcher's bill' of battle – like almost everything else we think we know about conflict – tells us very little about success, failure or even competence in command.

16. Strategy, Tactics and Friction

At its simplest, we might distinguish between the first two of these terms as 'strategy is what takes you to the battlefield, tactics are the tools employed once you get there'. That would be

Body of a soldier at Gettysburg, evidently killed by a shell. Casualties were almost eerily comparable, the usually accepted figures (including prisoners) being 23,055 Union, 23,231 Confederate. A draw then?

a crude, but not invalid, interpretation. One is entwined with, or even dependent on, the other and both are dependent on military and political policies. If a government has decided to spend a great deal of money on having a high-tech military establishment with the latest models of tanks, ships and aircraft but a relatively slender infantry force, the strategic and tactical practices of the high command of the military will be (or certainly ought to be) defined accordingly. Generals can only make war with the tools that they have. They may have a wide range of options depending on the flexibility and versatility of the forces under their command, but the simple fact is that you cannot fight a war at sea if you have no ships or deliver an airborne assault if you do not have the requisite aircraft.

Every strategic or tactical approach revolves around a combination of the available force, the nature of the environment and the intentions of the enemy. There are other considerations – cultural traditions, for example – but it would be extremely difficult to conduct successful operations without regard to those critical factors. Defining where strategy stops and tactics begin is a fruitless undertaking. The force on hand may not be well suited to the strategic goals identified by the political leadership, in which case the preferred strategic policies of the generals may have to be altered to take account of the nature and size of the force and its relationship to the countryside.

Equally, at a tactical level, the commanders on the ground may have to make radical changes to the standard operational practices in which the troops have been trained. In fact, even when a force undertakes exactly the sort of operations they have trained for in exactly the terrain in which they expected to have to fight, the reality of the battle is never identical to that envisaged by the policy-makers, the training staff or the designers and manufacturers of the arms. There is also the matter of dealing with the enemy, whose actions and practices may not turn out to be quite what the planning staff had in mind when the operation was put together.

As we have already seen, one of the well-worn phrases relating to war is that 'no plan survives contact with the enemy'. In fact, few plans really survive the first few hours of an operation even if the enemy takes no action at all. A deployment is prevented because it turns out that the tanks are too heavy for a particular bridge or too wide for a particular tunnel; a unit does not leave its start line at the correct moment because someone's watch has broken or radio communications are disrupted by an electrical storm or a batch of faulty batteries. It is not even the case that 'everything that can go wrong will go wrong'. There is no perfect scheme of analysis that can identify and make provision for every possible factor and possibility in conflict. In 2009, Donald Rumsfeld drew a barrage of criticism for a speech in which he referred to the challenges facing military intervention; he spoke of 'known knowns', 'known unknowns' and 'unknown unknowns'. He can be criticised legitimately for clumsy sentence structures, but his premise was perfectly sound. Commanders and analysts can identify the nature of clear threats and make allowances for threats that they know exist but whose nature is not clear, but in war there are often – if not always – threats (and opportunities) that will not become evident until operations start to have an impact on the situation.

Sometimes those threats will not become recognisable until after the operations have been concluded; they may never be recognised at all, but that does not mean that they did not have an influence on the development and outcome of operations. Even in the unimaginable situation where there is a perfect plan, where every member of the force is perfectly trained, perfectly equipped, perfectly briefed and in perfect sympathy with every aspect of the operations in hand, there would still be complications that cannot be predicted and pitfalls that cannot be avoided; it only takes one person to be 'having a bad day at the office' and some detail – however minor – will start to unravel. Any number of things that have proved totally reliable or predictable in the past might just happen to fail for the very first time at a crucial juncture.

Clausewitz christened these factors 'friction'; a very appropriate term. However carefully a force prepares for an operation, friction will have an impact on its conduct. Friction is, of course a normal part of existence; it is an everyday aspect of business activity whether caused by unexpected commercial considerations, ill-considered political initiatives, personality issues or traffic jams. The difference between business friction and military friction is that in business nobody gets shot because of it … though sometimes people get fired.

17. Offensive or Defensive?

Neither of these terms is quite as clear-cut as we might expect. An army which is, in general terms, pursuing a defensive policy may be obliged to make an attack to stabilise a situation; it may even have adopted a generally defensive strategy, but pursue an offensive policy at a tactical level on a day-to-day basis. An insurgency against an occupying power is, broadly speaking, strategically defensive; the enemy is already present and has possession of territory. The engagements, however, will tend to be offensive in nature: raids, sabotage and ambushes. The converse does not necessarily apply, or at least not universally across the theatre of operations. The occupiers will be obliged to protect the territory they have captured, but in order to defeat the insurgents they will have to mount offensive operations or surrender the initiative to their opponents.

This holds true in conventional conflicts as well; everyone is familiar with the concept of a counteroffensive. A counteroffensive may be mounted to regain an objective or territory or to distract the enemy and thus weaken his own offensive plans. It may be a means of reassuring the army or the public or the political class that the commanders are 'doing something' or it may be that the enemy's offensive has offered an opportunity to strike a significant blow. In short, it can be difficult, if not impossible, to really define what is an offensive or defensive action. This may be true at any level. The Crusades were undertaken to reclaim the Holy Land for Christianity. Everyone knows that, but is it really true? One could make a case that the Crusades were more to do with opening a second front against the expansion or resurgence of Islam in southern Europe, in which case the policy was defensive at the level of a grand international policy, but the actual campaigns were still offensive operations, both strategically and tactically.

18. Fronts, Flanks and the Rear

For once, these terms are both clear and valid; we even understand, without explanation, the expression 'turning a flank'. We know what they mean, but that does not prevent us making questionable assumptions. One of these is that a frontal attack is always crude and dangerous and with a risk of heavy casualties. There are situations where there is no other means of approaching the enemy. From late 1914 to the end of the First World War, the commanders on the Western Front had no option but to mount what were essentially frontal attacks because there simply were no flanks to turn. The same point can be made in regard to a position with men, weapons and fortifications pointing outward in all directions; which aspect is the flank? The attacking force may be able to detect a weak point in the structure and attack it, but that's not the same as finding a flank; any position can – and generally will – have a weak spot somewhere.

Commanders are acutely aware of the vulnerability of the flanks of their forces – for reasons which we will consider shortly – and therefore tend to deploy resources to ensure security. The risk, of course, is that they focus on the flanks and weaken the centre and then run the risk of having their army split into two fragments, each of which now has a powerful enemy force to its flank. We might come to the conclusion that a successful penetration of the enemy's front might be the most effective operational gambit. In addition to destabilising both portions of the enemy's force there will now be an opportunity to utterly disrupt or even destroy his communications in the very centre of his deployment area and perhaps even destroy his ability to continue the fight. The attacking commander may, of course, attempt to achieve a number of penetrations at different points in his opponent's line. The potential benefits are obvious. In addition to the possibility of disrupting the enemy in several locations simultaneously, one or more of his attacks may be no more than a feint to prevent the enemy from reallocating assets to face the real threat. Moreover, the attacker may choose to abandon or reinforce any of his attacks in the light of the developing battle. Alternatively, we could conclude that a penetration is actually the most risky of manoeuvres. If the defending force keeps its nerve – and can preserve its supply lines and operational cohesion – it may now have the opportunity to attack the flanks of the force that performed the penetration. If the penetrating force fails to destroy or repulse these counter-attacks, it may have gained nothing more than a bulge into the opposition's territory – a salient.

The salient may provide one or more useful starting points for a renewed offensive or it may turn out to be a white elephant. There is the risk that a counter-attack could sever the salient at the base and cut off large numbers of troops and there is also the danger that abandoning the salient could have an adverse effect on morale. Even if that is not the case, the salient may be more of a liability than an asset. The famous Operation Market Garden of 1944 only fell short of its objectives by a very small margin, and its designer, Field Marshal Montgomery, claimed that it had been at least 75 per cent successful. In a very limited sense, this was true. The forces did acquire almost all of the objectives; however, failure to take the final bridge at Arnhem did not just mean that Montgomery failed to acquire the highway through the Netherlands and over the Rhine, which he believed would greatly hasten the collapse of Nazi Germany and bring the war to a close. Having delivered what he called a 'pencil-like thrust', he was left with a salient stretching for 60 miles into enemy territory but going nowhere in particular. The entire length of the salient was therefore vulnerable to counter-attacks, but worse than that, the counter-attacks did not really have to be successful in the sense of eliminating the salient; they just had to be enough to keep the Allied forces tied down in defending it.

If frontal attacks can have profound weaknesses, is it necessarily advantageous to attack on a flank? There are certainly strong arguments to support such manoeuvres. An attack to the flank can bring some of the same advantages as a successful penetration, and possibly at less risk. Defending troops tend – understandably – to become concerned when they are made aware that the enemy is operating to the left or right of their positions as well as to the front. They are naturally worried that they may have to face attacks from two directions initially, and potentially from the rear as well, if the enemy's flanking manoeuvre is completely successful. They may even become surrounded and run out of food and ammunition. A flank attack carries less risk than an attempt at penetration for several reasons. If the attack is not successful there is a better chance of disengaging successfully and with limited losses if the enemy is concentrated to the front of the attacker, whereas in a penetration the enemy is to be found on either side and to the front as well.

Assuming that there are no significant enemy forces in the vicinity, there is little risk of the flanking troops being outflanked in turn or being attacked in the rear and finding themselves surrounded. The potential benefits of a successful flank attack can be just as great as those which might come from a penetration in the front of the enemy's line, but the action of mounting an attack may actually force a break in the attackers' lines which can be exploited by the defender. These are concepts which are difficult to explain in words, but easy to demonstrate in diagrams.

In order to see the positive and negative aspects of any of them, it is useful to look at the concepts of interior and exterior lines. Once again, the ideas are simple, almost elegantly so, but the practice is challenging and by no means constant. Like the issues of 'defensive and offensive' or the concepts of 'strategic and tactical', their value – even sometimes their existence – is not easy to define.

As a simple example, we can look at a fortress or a near-circular position. The forces of the defending commander (A) have 'interior lines' – that is to say they are in a compact position relative to the enemy. Moving men or material from one part of the front to another is a fairly simple business. The lines of communication across the fortress are as secure as they can be and the path from one part of the front to another is short and therefore can be travelled quickly. Any action by the enemy commander (B) can be met with a rapid response. Clearly that is a situation with positive aspects.

The enemy surrounding the position is certainly at a disadvantage in certain respects. If (B) wishes to move resources from one point to another it will take longer than for the opposition to match his movements; moreover, any adjustments to his deployment may be observed more easily. (A) on the other hand – with his interior lines – may well be able to move forces around the battle front in complete secrecy. If both (A) and (B) are to have the same density of troops all along their respective fronts, (B)

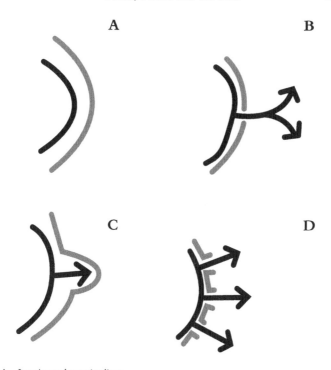

A – Interior and exterior lines.
B – A successful penetration disrupting the enemy's position.
C – An unsuccessful penetration attempt resulting in a salient.
D – Multiple penetrations leading to extreme disruption.

will have to have more men at his disposal since his front line will be the longer. In practice, (B) will have to have a great many more troops if he is to be secure in his position. Since (A) can concentrate his forces much more quickly, he can achieve local superiority at any given point of the battlefield.

The advantages of interior lines are considerable, but there is no such thing as a perfect approach to deployment. A real risk to very tight interior lines is that the enemy may be able to surround the position, and, in the case of a fortress under siege, will almost inevitably have done so and can simply starve the defenders out of action. Moreover, if the position is very tightly constrained, and therefore has a very high density of soldiers, the force inside it is likely to be very vulnerable to missile fire. The target area is easily identified and the chances of inflicting damage with each and every round are greatly increased. This may be offset to some degree by ensuring that all of the defending troops are well protected in bunkers or trenches, but that in turn may reduce their ability to react to developments or even to see or fire at the attackers.

It is, of course, rare for interior and exterior lines to be quite so starkly defined. A wider dispersal of force gives certain advantages to the army deployed on exterior lines – primarily a wider choice of approaches for attacks and greater opportunities for manoeuvre outwith the opposition's field of fire or observation. Exterior lines may also offer a better range of options in terms of wider manoeuvres: the ability to take advantage of features in the landscape and outflank an opponent or even cut his lines of communication entirely. When we examine a local tactical situation or the overall relationship on a strategic level, it is generally fairly simple to identify which of the two forces has established itself on interior or exterior lines, but it is often the case that each will have adopted, or have been obliged to adopt, interior or exterior lines deployments as a reaction to the deployments of the enemy, to facilitate planned initiatives or possibly because of features in the terrain. This applies at all levels, rather like the distinction between offensive and defensive. The posture of the army, or a corps, division, brigade or battalion, may be 'interior', but one or more of the constituent parts of the force can be 'exterior'. A salient, whether it contains one rifle section or an entire corps, is always an example of 'interior lines' since it must self-evidently be surrounded by the enemy on three sides. However, the main body of the force, whether it is two sections of a platoon or the balance of a massive group of army corps, may well be arrayed on 'exterior lines'.

19. Morale and Confidence

It is not unreasonable to ask whether these are really two separate characteristics. A high level in one aspect certainly has an impact on the other, but is important to distinguish between either (or both) and competence, and to remember that even confident forces with a high level of ability are never guaranteed to have success in battle. History is replete with examples of whole armies and individual units which had every confidence in their ability to wage war right up to the moment of their utter and complete defeat and, of course, there have been innumerable generals who really thought they were more than capable of inflicting defeat on the enemy, only to find themselves on the receiving end of a thorough drubbing.

Strong morale and confidence has to be balanced with realism and reinforced with consistent success if it is to be maintained in any circumstance other than clear victory. If the morale is genuinely strong and based in a genuine and merited level of confidence in the ability of the army, even quite dramatic set-backs can be weathered. On the other hand, if contact with the enemy reveals a lack of competence among the troops or the leadership, morale is very likely to plummet.

None of this will come as any surprise to even the most casual student of military history; the question is 'what to do about it'. There are many tools that a commander can use to raise or maintain the morale of the troops. This is the case with everything else in war. The actions themselves are simple in principle, but often difficult to achieve. Good training which is – or at least seems to be – relevant to the business of soldiering is clearly a vital component. The troops need to have confidence their officers, and have a belief that they are being led by competent people who are serious about the task in hand and are utterly committed to the well-being of their subordinates.

20. Structures and Articulation

Even the most talented and well-trained commander can only relate directly to a relatively small group of people. A force as small as one hundred men must have an internal structure; the commander can hardly issue orders to each individual under his command. The force must therefore have a command structure, one that allows varying portions of the force to be committed to specific tasks within the overall plan of action. As the forces become larger, the structure requires more levels of management and a variety of types of departments if it is to function at all, let alone to an adequate standard. The nature of the structure of armies has had a certain degree of stability over the centuries, though naturally, as armies have become larger and have come to cover a much greater acreage, it has become necessary to have more command levels. Equally, as the operational demands of armies have become more sophisticated there has been a massive expansion in the number and size of the ancillary elements – more technicians and engineers in an ever-increasing number of subdisciplines – but in the combat units, the number of men under command at each level has changed very little since classical times. The classic Roman structure of centuries and cohorts is not a million miles away from the average size of modern companies and battalions.

The names of the units have changed, but the level of responsibility of commanders has remained fairly static save for the development of the largest bodies, what Second World War officers referred to as an 'Army Group'.

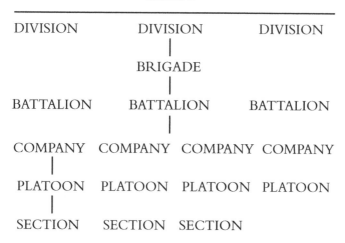

At each level of command there will inevitably be a headquarters element within the unit or formation structure. The corps headquarters will include a veritable raft of administration and intelligence and other specialist personnel which is mirrored on a smaller scale at the divisional and brigade level, and that will generally be augmented with artillery, armour, aviation and other units, and will be deployed in support of the constituent parts of the formation, as required. A similar structure may be – and generally is – present within the battalion as well. The headquarters of the battalion will have administrative and tactical responsibility for specialist subunits such as a mortar or anti-tank platoon, not to mention a handful of combat engineers, medical staff, a detachment of military police, an intelligence section and possibly many others.

The precise structure does, of course, vary from time to time and from one army to another and – especially for the higher formations – it may be adjusted to reflect a given tactical or strategic environment or as a product of the demands of the battlefield.

There is any number of examples of a division being removed from its parent corps to temporarily reinforce another. Although that is part of the value of a system of articulation, it has weaknesses when taken to extremes or when it is applied for long periods. The headquarters of a brigade or a division can cope with the additional administrative workload of one more unit under command for a short period, but not indefinitely, since the structure of the headquarters, and the manpower and transport facilities allocated to it, were designed to cope with a given number of units – three, more often than not – in the first place.

In every army in the world, the smallest combat organism within the army consists of a mere handful of people. There are instances where a single soldier or a team of two is deployed to a task, such as a sentry or a sniper, but generally the starting point is the infantry squad (or in the British Army, the section) or the crew of a vehicle or weapon. In most armies the squad has about ten members. It is not a hard and fast rule; armies have experimented with varying squad strengths from as low as six to as many as fourteen, but nine or ten is the most common nominal strength. Subunits of around this size have a long history; the Roman *optio* had responsibility for eight soldiers. Corporals have a greater command responsibility than any other combat leader. They have to organise, lead and control eight or nine men directly and continuously. The commander of a regiment (or brigade in the British tradition) of three 700-man battalions will generally only have to deal directly with three battalion commanders and perhaps a couple of specialist officers at his daily briefings and intermittently throughout the day. Most of the time, the battalion commanders and the specialists will (or certainly should) be getting on with their tasks with little or no input from the brigade commander until such time as a new situation arises which requires his intervention, or the current operation is concluded and he needs to present his plans for the next one.

At every tier of the structure above the squad, the officer in charge mostly has three or four subordinates to deal with: the leaders of three sections to a platoon, three or four platoons to the company, three, four or five companies in the battalion, three battalions in the regiment or brigade and three brigades serving together to form a division. The size and composition of formations at all levels has varied at different times and for different applications and is also affected by the availability of manpower. If numbers are reduced, perhaps by casualties or desertions, a point is reached where constituent parts might be amalgamated to provide units of a viable strength or that are practical to administer. This is a recurring issue for small-unit commanders, especially in a prolonged or particularly intense conflict when casualties are high but replacements are not readily available. Is it more useful to have the flexibility of three weaker platoons or the greater firepower of two strong ones? There is not a simple answer; the commander must make a decision based on what he knows of his troops and his situation and what he can deduce about the immediate future.

The structure of the force is, therefore, the vehicle for its articulation; that is to say, the means of general organisation and for the application of force. At the most basic level, it allows the commander – at whatever level – to dedicate a portion of his force to a specific task or occupation.

We can see a parallel to this when we look at shift work in commerce. The 'shift on' will generally require a range of skills and a range of management personnel to ensure a steady quality and quantity of output and to take action over any issues which may arise in the course of the shift – workplace injuries, machine failures, the discovery that there is not an adequate quantity or quality of materials. It is not vital that the managing director is on hand at every moment of the night and day, but it is vital that there is someone on hand who can make immediate judgements about problems that arise. Most of these decisions will be fairly straightforward, some will be very challenging, a small number will require the person in the hot seat to phone the managing director at 3 a.m. and risk the ire of someone disturbed in the middle of the night.

Articulation is what enables commanders to apply their force from different directions or in different roles, and we can see this sort of thing depicted in films. For example, a commander is briefing his subordinates and tells them, 'A Company will attack here, B Company over there, C Company will provide covering fire and D Company will pass around the flank of the enemy and attack him in the rear.'

Articulation is, however, much more than just a matter of battlefield orders. It is what allows training at all the different command levels in the unit – training by sections, platoons and companies within the battalion, training of battalions so that they co-operate effectively within the brigade and training in brigade exercises so that commanders and their staffs can learn to ensure that their units function well within the division.

An important aspect of developing an effective and universal system of articulation throughout the army is to bring about compatibility. A unit that has had very heavy casualties or has to undertake a very challenging task may need to be augmented by elements – or even the entirety – of another unit. When we read of this or that unit (X) carrying out an operation with 'a company of unit (Y) under command', the meaning is fairly clear, but the process could not be managed effectively if there was not a standard organisational structure that applied to both (X) and (Y).

21. Building a Better Mousetrap

The compatibility and interaction that is so desirable for effective articulation is also critical in armament. Although most forces have a positively bewildering array of weaponry, each weapon has one or more functions that are more or less system-specific. This or that weapon may have a variety of functions and may share that function with more than one part of the arsenal, but there will generally be one application for which that weapon is more effective than others. All of them share a common purpose in the sense that they are designed to inflict casualties or in some cases – smoke, for example – to prevent the enemy from doing so. They share another characteristic; there is a perpetual pressure for improvements in our capacity to kill or maim. The introduction of new weapons and/or tactical approaches is seldom a steady or predictable process.

An issue that hampered the development of the machine gun was the development of highly reliable bolt-action rifles with a good rate of fire. There was a case to be made that well-trained riflemen were more than a match for unreliable machine guns. There were two rather obvious weaknesses to the 'pro-rifleman' argument. One was that it takes a long time to train really proficient infantry soldiers, and if they are to be brought to a degree of excellence they have to be the best material. A massive army of conscripts will have a very tiny proportion of men who are sufficiently fit and motivated, compared to a small professional army.

The other issue was technology. The machine guns of the 1870s really were unreliable; they were also cumbersome and had to be drawn by horses. They were closer to artillery than to small arms, but the technology was new and therefore likely to be the subject of rapid improvement. The machine gun became more reliable and lighter, which meant it could be applied differently. The result was the trench warfare of 1914–18. It was still true that a skilful section of riflemen could outmanoeuvre and destroy an isolated machine-gun team, but there were not that many genuinely proficient infantrymen. Widespread use of machine guns meant that infantry units were pinned down and dug themselves into the earth for safety. As long as the infantry kept their heads down in a trench they were impervious to machine-gun fire, so if the enemy wanted to get them out of a position he would have to apply artillery. Within a short time, it became all too apparent that well-prepared trenches were a good deal less vulnerable to artillery than anyone would have thought and that the massive barrages before attacks often achieved little apart from tearing up the terrain which, in turn, made it more difficult for the infantry to conduct the assault.

In due course, the conditions that developed on the Western Front from 1914 onward led to the development of the tank. The concept was not new; Leonardo Da Vinci had given thought to an armoured vehicle centuries before, but there was no means of propelling one. The weight of armour and soldiers would have been far too great for a team of horses, and the same applied to a variety of steam-driven designs in the nineteenth century. The internal combustion engine made the tank a viable proposition, but tanks are expensive and cannot operate everywhere, so the machine gun problem still exists, but has been circumvented to some extent by issuing every soldier with an automatic weapon.

The 'better mousetrap' process can be seen in every facet of military activity. It is not always effective. Some years ago, a squad of American troops was kept pinned down all day by two insurgent fighters armed with 70-year-old bolt-action rifles at a range of about 700 yards. No one was killed, but the Americans were prevented from carrying out their allotted task for a whole day by two men with antique weapons and a handful of bullets.

Advances in weaponry are important, but their value can be exaggerated. There is a form of the law of diminishing returns in military spending and development. A new model of fighter aircraft costing £100 million is only likely to be a very marginal improvement over its predecessor or over another aircraft at half the price. The application of force and the quality of pilot training may be much more of an issue. Recent (2013) exercises in southern Europe indicated that well-trained Turkish pilots in 40-year-old Phantoms were quite capable of giving a very hard time to opponents in much more modern aircraft. Evidently, success in combat is not simply a matter of technology, but also of skill and application.

Effective application of force can also be affected by issues that seem painfully obvious once they have come to light. For nearly as long as aircraft have carried machine guns, it has been a common practice to load a high proportion of 'tracer' bullets – ones which burn phosphorus to emit a great deal of light so that the pilot can see where his bullets are going, since a standard bullet cannot possibly been seen in flight. This made perfectly good sense for decades until quite recently when it occurred to someone that a tracer round does not behave like a normal one.

A better mousetrap? The Tiger tank certainly struck fear into the heart of the enemy – they were constantly being spotted in Normandy when they simply were not there. But they were complex, heavy and prone to breaking down. At Kursk, their transmissions were found to be fragile. They constituted only 5 per cent of the Panzers and assault guns utilised by the Ninth Army, Fourth Panzer Army, and Army Detachment Kempf in the battle. Yet their psychological impact is confirmed by the triumphant summing up of the clash in the famous *Izvestia* headline, 'The Tigers are burning'. (*Zitadelle*)

It is very much hotter, which alters the trajectory and it loses a good deal of its weight in flight, which changes the trajectory to an even greater extent. In essence, if the tracer rounds are 'on target', the pilot may not know exactly where the other rounds are going, but he or she should be confident that they are not scoring hits on the enemy that they are engaging. This would seem to be pretty straightforward but, like most simple things, it is only simple once somebody explains it.

Having a better mousetrap is highly desirable, but how do you know if the mousetrap is genuinely better as opposed to just being new, and how do you evaluate whether a marginal increase in utility is justified by an enormous increase in cost? The only really valuable answer to these questions is utility on the battle-field. With the best will in the world, no amount of training and no system of exercises will ever really equate to the experience of battle. In the world of rock and roll, this is known as 'suck it and see'; you never really know if a thing will do what it is supposed to do.

22. Getting it Right and Getting it Wrong

For the military, success is the only real cause in warfare. That is not absolutely the case in the political arena. For the political class, success can be defined in various ways even in the face of defeat in battle, such as showing support for an ally or demonstrating an ideological or religious position.

For the military, it has to be about winning. Naturally it is easier to win by taking sensible courses of action, though it is only fair to recognise that sometimes conflicts are won despite the quality of commanders and not because of it. Equally, success in war can be the product of several factors that are not brought about by either competence or incompetence of the commanders on one side or the other. This need not have any impact on their reputation in history. George Washington was not what one would call a good general, or even a competent one, but his fame is assured because he was on the winning side and because he became president. The invention of tales about felling cherry trees and not lying about it has not done his legacy any harm either.

Commanders do not set out to fail. Sometimes they are obliged to fight when they would rather not, but they do not go looking for defeat; even in the darkest times, if a commander offers or accepts battle, he does his best to win.

However well an operation is planned and however carefully it is rehearsed, it may still fail – even if the enemy is not prepared for battle and even if it appears to have been a major victory at the time. The Imperial Japanese Navy's strike at Pearl Harbor in December 1941 was a masterpiece of naval warfare. The Japanese fleet managed to avoid detection on a long voyage and delivered its attack with skill, undertaking a daring and intelligent task. The Americans were not ready for war in any sense and several of their prized battleships – still seen as the backbone of war at sea – were sunk or very badly damaged.

The Japanese fleet was able to conduct the operation and withdraw with marginal losses, but the operation was still a failure. Four American aircraft carriers – *Hornet*, *Yorktown*, *Ranger* and *Wasp* were in the Atlantic, either training or effectively giving unofficial aid to Britain. The two carriers in the Pacific – *Lexington* and *Enterprise* – had been delivering aircraft to marine bases at Midway and Wake, respectively, and thereby escaped being bombed or torpedoed at Pearl. Failure to catch any of the American carriers was an immense blow, but it may not have seemed quite so crucial at the time. Naval thinking still saw carrier-borne aircraft as a means of locating the enemy's battleships and then causing enough damage to prevent their escape before friendly ones could get close enough to engage and finish the job. The principle seemed to have been validated by the British success against the German *Bismarck*, which had been sunk on its one and only operational voyage just six months before. Aircraft (and obsolete ones at that) from the British carrier *Ark Royal* had found *Bismarck* and damaged her sufficiently for the surface vessels to close in and destroy her. The progress of the war would demonstrate the huge significance on the carrier, but that was not apparent in December 1941, so although Pearl Harbor would eventually turn out to have been a failure, it was, initially, seen as a great victory.

The daring nature of the Pearl Harbor raid has rather overshadowed the other operations that were part of the same strategy. In part, this is due to the sheer impact of a dramatic surprise attack; to a lesser degree the difference in time zones and the International Date Line have made a series of initiatives look like they cover a period of two days rather than a matter of a few hours. Although the Japanese military was already heavily engaged in China, within a few hours they managed to mount major offensive operations at Hawaii, the northeast coast of Malaysia, the Netherlands East Indies (Indonesia), French Indo-China (Vietnam), Hong Kong and Wake Island in the middle of the Pacific Ocean.

All of these offensives were successful and the pattern was set for expansion to provide a gigantic defensive perimeter which could be reinforced by sea. The rationale was not far removed from the policies of two of the countries that the Japanese had attacked; Britain and the Netherlands. Both had concluded that any threat to their Far East colonies could be met in the short term by forces stationed 'in country', which would hold an invader at bay until the arrival or reinforcements from home. Japanese planners had felt that the greatly enlarged Empire would be secure because of its interior lines and because the Americans would be so crippled by the Pearl Harbor attack and the British so stretched by their war with the Germans and Italians in the Atlantic and Mediterranean that there would be ample time to build impregnable defences. In fact, all they had done was set up a vast boundary which was far too long to be securely defended; precisely the problem that had faced the British in 1941 and had afforded the Japanese the opportunity to make their daring and initially successful offensives throughout the Far East and the Pacific.

23. Deterrence and Appeasement

The term deterrence is almost universally applied to nuclear weapons. The great problem with having a nuclear arsenal is that it is very difficult to get rid of it. The expense is enormous, especially so if the delivery method is to be by missile from submarines. In order to justify that expense to the public, a government has to continually argue that the deterrent is absolutely vital to the defence of the nation and to the strategic role that the nation might want to play on the world stage. The latter issue is tied into how the political class perceives the diplomatic importance of the nation and consequently their view of themselves as world-class statesmen. In the case of the United Kingdom, that perception is allied to – if not dependent on – a view that holds Britain to have a 'special relationship' which makes the UK the most important and influential ally of the United States, and on the allegedly vital necessity of retaining Britain's position as a permanent member of the United Nations Security Council. Both of these perceptions are products of the British role in defeating the Axis powers in the Second World War, and both are rather questionable. When the UN was a new body, it was crucial that all of the great powers were active in its deliberations and it really would have been unthinkable that Britain was not at the 'top table'. The world does move on, however, and the environment of the twenty-first century is rather different from that of 1945. The UN itself has recognised this in a roundabout way, though it took more than twenty years. China was one of the five original permanent members of the Security Council, but by 1949 China's communist party under Mao had driven the national government under Chiang Kai-Shek off the mainland. For the next two decades, the Taiwan government continued to provide a permanent representative to the council despite being a power of absolutely no consequence whatsoever, while mainland China was not represented at the UN at all. In 1945, Britain still had imperial and colonial interests all over the world and was a major military power in every sense. Little more than a decade later, the Suez Crisis demonstrated that Britain was no longer capable of taking on a major operation – even with the support of France and Israel – without the approval of the United States.

For the benefit of their domestic audiences, the British and French governments continued for many years to consider themselves members of a 'big four' group of superpowers alongside America and the Soviet Union. In fact, they sometimes still present that belief – but in reality it is abundantly clear that neither London nor Paris is quite so influential as all that.

The loss of Empire and the diminution of Britain and France as conventional military powers, and a general reduction in their international status, led both countries to develop nuclear weapons of their own. Neither was capable of defending itself unaided against the kind of nuclear onslaught that the Cold War presented. Their arsenals could have inflicted gigantic damage on the Soviet Union, but they were not sufficient to mount a 'first strike' that would be powerful enough to prevent an even more massive retaliation which really would have totally obliterated France or Britain for generations – perhaps forever.

Although there were many factors propelling both governments toward adopting a nuclear capability, over the decades,
the single most significant one has been prestige. Having 'the
bomb' has been an important lever in preserving the permanent
seat on the UN security council and in having a voice – however marginal in practice – in negotiations with what used to
be the Soviet Union. Successive British governments – and
French ones too – have described 'the bomb' as an 'independent
nuclear deterrent', a phrase that demands a little critical examination. It would be politically impossible to deploy the weapons
without the approval of the United States – in fact, it would be
fairly difficult to imagine a scenario in which their use would be
acceptable to the British people. It might have been acceptable as
a response to a Soviet strike in the 1960s or 1970s, but it would
have been most unlikely that the Soviets would have struck at all,
and even less so that they might mount a strike in Europe but
not against America, in which case the British and French 'contribution' would have been of marginal importance. The term
'independent' is questionable at best and just plain dishonest at
worst. 'Deterrent' is equally questionable.

Since long before the fall of the Soviet Union, the threat of
a weapon of mass destruction (not a very useful term, but one
that is widely used and well understood) has not been a missile
war with a great superpower, but an attack by a 'rogue state' or
even possibly a non-governmental body. In either case, the sort
of people who would arrange the detonation of a nuclear device
are not the sort of people who would be deterred by the threat
of a massively destructive response – even if a viable target could
be identified.

It is not at all difficult to find politicians of all British parties who – privately – would like to see the decommissioning
of Trident and even more who would rather it was not replaced
when it reaches the end of its operational life, but both of the
main parties believe that there would be a cost in credibility.

Failure to replace Trident could be seen as an admission that it had been, at best, of dubious value for some time, and both parties have not only invested a great deal of political capital, but have also spent a great deal of money which might otherwise have been applied to more useful projects.

Deterrence, however, is a wider matter than just the avoidance of a nuclear exchange; it may be employed, and be effective, on a much less cataclysmic scale. The presence of even a small military force from an allied nation can be enough to make invasion by a larger and more powerful neighbour an unattractive proposition. The United Kingdom retained a force in Belize for a decade after independence, which helped to deter an invasion by Guatemala. Continuing American bases in South Korea fulfil the same function. At less than 30,000 personnel (in 2012), the extent of the deployment is not so great that it would have any immediate impact on a North Korean invasion, but it is enough to demonstrate that the United States would take steps to defeat an attack.

Appeasement could be construed as a direct alternative to deterrence. If one side chooses to accede to the will of a hostile neighbour in the hope of averting conflict, that is certainly appeasement, though experience would suggest that accepting demands will only lead to another set of demands being made, and then another, and then another. It may, however, be no more than a stop-gap measure, a means of keeping the enemy at arm's length for a period which can enable the defender to develop his own military capacity. Neville Chamberlain's actions in 1938 can certainly be seen in that light. Regardless of whether he actually believed that Hitler could really be dissuaded from further expansions of the Third Reich, the British government did take steps – though perhaps inadequate ones – to prepare for war. In that respect, surrender is not always radically different in effect. The settlement which one side sees as evidence of victory may be seen by the defeated party as no more than a temporary lull in the struggle while they regroup and re-equip for another round of hostilities.

All three of these concepts – deterrence, appeasement and surrender – can be reflected at a strategic and tactical level as well as a political one. The occupation of a very strong position or the recognised (or even imagined) superiority of force in a given location may be enough to deter an opponent from mounting operations there. Similarly, abandoning a particular position or operation may encourage the enemy to believe that he has gained as much as he can at that place and at that time, and give the defender an opportunity to rationalise his own deployments, gather reinforcements or receive supplies.

Surrender can be a justifiable option – or even the only practical option – in certain circumstances. If there is no means of attaining victory there may be no point in further bloodshed; for most people and in most circumstances, life in a prison camp is going to be preferable to death on the battlefield. The problem with a local surrender is, of course, the effect that it might have on other parts of the battlefield or in other theatres. So long as a beleaguered force is under arms, the enemy will have to devote resources to keeping them contained and in due course destroyed, so an isolated force which has no real goals of its own is still an asset in the wider tactical picture.

A strategic surrender has greater implications for the political class as well as the military hierarchy. Defeat on the battlefield will undermine confidence among the people and may lead to occupation. On the other hand, a negotiated surrender may allow the political establishment to retain its authority and preserve whatever has survived in the way of military forces as the kernel of a new army which might, in the fullness of time, allow the recovery of whatever losses were suffered by surrender.

24. When Everything Goes Wrong

It is perfectly possible to develop a strategy and tactical approach which seem to be sound and promising, or at least to offer promising prospects, and yet fail completely. In the summer of 1863, Robert E. Lee embarked on an invasion of Federal territory. A previous attempt to carry the war to the enemy had not been especially successful, but Lee could be forgiven for concluding that the circumstances had changed or that mounting an offensive was his only viable course of action. He had total confidence in his army, though perhaps rather less in his subordinates. He had recently reorganised the Army of Northern Virginia into three corps of equal size under Generals Longstreet, Hill and Ewell.

All of these were competent and proven officers, but Hill and Ewell were new to this level of command. Their promotion is actually an indication that he did not have complete confidence in either of them. Until this point the army had, essentially, operated as two wings; one under Longstreet and the other under Jackson, who had recently died of wounds. The new structure did offer an improved degree of flexibility, but also suggests that Lee was not sufficiently confident of either Hill or Ewell to entrust them with half of the army.

The attractions of mounting an offensive were considerable. Successful – though costly – operations had prevented the Union forces from capturing the Confederate capital at Richmond, but the war had bogged down into a stalemate. The Confederacy was running out of steam and could not maintain a war indefinitely, so a new initiative was required. Taking the army into

Maryland and Pennsylvania would allow Lee to feed his army at the expense of the enemy and give the farmers of Virginia some relief. More importantly, one or two sharp victories on Union soil might have an assortment of desirable outcomes. Success in battle could encourage anti-war sentiment in the North, it could cripple the Union army in the Eastern Theatre and force Lincoln to divert forces from elsewhere, and the campaign might even bring the Confederacy diplomatic recognition from Britain and France. It might even be possible to seize Washington itself and force Lincoln to accept some sort of peace.

The campaign did not start badly. Lee was able to extricate his army and steal a march on General Hooker and was thus able to move through Maryland and Pennsylvania unopposed. What he could not do was keep his army concentrated; if the troops were to be fed, they had to be able to forage far and wide. Hooker got his own army on the move with creditable speed and correctly guessed Lee's general intentions, but his failure to prevent Lee's initiative led to him being sacked and replaced by General Meade. Meade continued with Hooker's plans and the Army of the Potomac headed north with commendable speed and efficiency to contain Lee and prevent him from capturing a major federal city such as Baltimore or Philadelphia, but things had already started to come unstuck for Lee. His cavalry commander, General Stewart, had embarked on a major, but rather pointless, adventure through Union territory, which meant that he failed to perform the primary task of his force, which was to monitor the movements of the enemy and prevent him from gathering intelligence about the movement of the main body of the Confederate army. Lack of information from Stewart meant that Lee found himself confronted by a rapidly concentrating Union force of great strength which might well do more than block his path to Washington, but it might prevent him returning to Confederate territory.

As soon as the threat became apparent, Lee ordered a concentration of his army, though that was less effective than it might

Bird's-eye lithograph of the Gettysburg battlefield, published soon after the event.

have been due to his failure to make the rendezvous absolutely clear; some officers were not sure if they were to march on Gettysburg or Cashtown. As it turned out, the decision was made for him and he was drawn into a battle that he had not chosen when a division of Hill's corp under General Heth became embroiled in an action with Union cavalry forces at Gettysburg.

Although the town was of no consequence, it was at the junction of several roads. That was advantageous for Lee in that his army was spread around the area and there were adequate roads headed from all points of the compass which would allow him to concentrate his assets at Gettysburg, but it was just as useful, if not more so, to Meade for exactly the same reason. Lee believed that he would be able to concentrate more quickly than Meade and, although he had wanted to fight elsewhere, the early part of the battle – after some very stiff fighting – seemed to support his conclusion; however, he had lost sight of the basic premise of his own plan of campaign. He was not opposed to fighting a battle but he had intended to conduct an offensive strategy with defensive tactics: to enter the enemy's territory and find a good defensive position which Meade would have to attack. Instead of that, he now found that his army, though concentrated in the strictest sense, was distributed to three points of the compass around a strong position.

Instead of being able to force the enemy to give battle in dribs and drabs as they arrived on the battlefield and thus enjoy the advantage of local superiority on terrain of his choice, he had been obliged to adopt a wide front to ensure that Meade could not outflank him, get behind the Confederate army, and cut their lines of communication.

Meade's forces had been driven back through the town in the first day of fighting, but had recovered their composure, largely through the failure of General Ewell to pursue the retreating Union troops with vigour. By the end of the day he was now in a very useful position. The balance of his army would arrive along roads that converged in the rear of that position and were thus protected from Confederate interference unless Lee mounted a major attack over terrain that favoured the defence. The dispositions of the Confederate army before the battle and the road network gave Meade a major advantage. As time passed, his army would become stronger by the hour and his forces would have interior lines, whereas by the forenoon of the second day of the battle most of Lee's army had already arrived, but were spread over a much longer front.

At first glance, the day had gone well for Lee, but failure to bring about a complete victory and drive the Federal army into a precipitate retreat meant that he was now in a poor position, with an increasingly powerful enemy force in front of him which had to be defeated if his strategic plan was to bear fruit. More than that, if he did not fight Meade and attempted to withdraw he would run the risks of a retreat of one hundred miles or more with the enemy in pursuit. Lee decided to continue the attack for a second day. A demonstration by Ewell on the Federal right flank, which was intended to prevent Meade from countering a move on the left by Longstreet, did not have very much effect and Longstreet's own attack was not mounted until late in

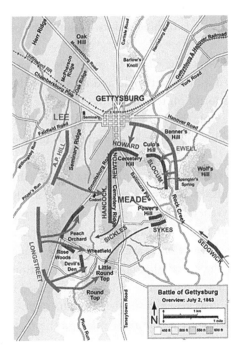

Second day; Longstreet fails to roll up the Federal left flank..

the day and failed to overrun the Federal position as intended. As time passed, Meade continued to receive additional troops and by mid-morning on the third day he had amassed superior numbers, though this was probably not clear to Lee. Stewart had arrived with the cavalry, but too late to make much difference to the situation; however, Lee was, by this time, adamant that the battle should continue.

Lee's rationale was not without merit. Withdrawal and redeployment to a better position which might prompt Meade to attack him would have been difficult to achieve and involved the risk that his army might be disrupted and defeated in a series of smaller actions over which he would only have a very limited degree of control. He could not be everywhere at once, and the two ends of his army were several miles apart. Moreover, there was no guarantee that adopting a defensive position would force Meade to fight on Lee's terms. He might simply redeploy to contain the Confederates and wait for their supplies to run out, at which point Lee would be forced to disengage in the face of an enemy force which was getting stronger by the day. Furthermore, a withdrawal in the face of the enemy might have a damaging effect on morale.

Having tried and failed to turn the Federal flanks, Lee now turned his attention to the centre. If he could break Meade's line in the middle of the battlefield he would effectively split the Federal army into two and – so he hoped – enable him to defeat one or other part of it, in which case the remaining portion might well retreat in disorder. Lee has been widely criticised for his decision and for the execution of his plan, but it is important to remember that his experience of fighting the Federal army would have suggested that this was a distinct possibility; similar initiatives had worked in the past. Every student of the Civil War will be aware that the assault – widely known as 'Pickett's Charge' though he was only one of three divisional commanders, and General Longstreet had planned the actual attack – was a complete failure. The attack was mounted some hours after Lee had ordered and so subsidiary operations on the Federal left flank were utterly compromised and various units which should have moved in to support the attack once it was under way failed to do so. It is not clear that either of these factors, or several others relating to the performance of the initial Confederate artillery barrage, made any real difference, though they have provided topics for heated discussion over the last 150 years.

Some of the factors that brought about Lee's defeat are clear, but it is interesting that the battle is always seen as 'Lee's defeat' rather than 'Meade's victory'. So it is worth giving some thought to the Federal perspective.

During the day before the initial action, a Federal cavalry commander, General Buford, located the enemy and found a strong position where he could block their progress the following day with the leading elements of the Federal infantry. By doing so, he forced a battle that the enemy did not want. As the balance of the Federal army concentrated forward to the Gettysburg area, some Federal officers planned to retreat before Lee's army, but General Hancock saw an opportunity, took control of the situation (Meade was still travelling northwards), and deployed for a defensive action, confident that Lee would accept battle on those terms.

Gettysburg is very much Hancock's battle in the sense that by the time Meade arrived on the scene the major decisions had been made. That does not reflect badly on Meade; quite the reverse. He chose to trust the judgement of an officer whom he knew to be a competent and conscientious soldier. Aware of the abilities and limitations of his army, Meade chose to adopt the sensible course of taking advantage of the terrain and waiting for the enemy to come to him. He was fighting the battle of his choosing, not Lee's, which is a hallmark of competent generalship. Over the next two days, he kept his composure, reacting to the movements of his enemy in a calm and rational manner. Once a victory had been secured, he did not over-reach and imperil his success.

Meade has been the subject of endless criticism for not pursuing Lee, but letting him withdraw to Virginia. That course of action might have led to even greater damage to the Confederate army, but equally it might not. Meade's army was exhausted. Some units had been on the battlefield for three days, many had suffered dreadful casualties and all of them had had several days of rapid forced marches, but they had just defeated an enemy who was widely seen as invincible. They may have been tired beyond measure, but they had established a moral superiority; they would no longer be afraid of the Army of Northern Virginia and the rebel yell. The advantage in confidence and morale gained at Gettysburg might well be lost in a series of actions against determined Confederate rearguards, and although Meade had a pretty good idea of the strength of the Confederate army, he could not have had very much information about the extent of their losses beyond a guess that they would not be dissimilar to his own, which were very considerable. It might easily happen that he could mount a pursuit, only to have a leading division or even a corps routed or destroyed in a battle somewhere along the road from Gettysburg to the Potomac River.

This one battle – and there are many others that would have served the purpose – demonstrates a wide variety of issues. Two of Lee's immediate subordinates were not yet comfortable in their new status as corps commanders and the third, Longstreet, was not happy with the plan. Hesitation on the part of all three may not have brought defeat but it did not help. Lee's concentration of the army was not as rapid or effective as it might have been, allowing Meade's subordinates to gather enough force – though it was touch and go at different junctures – to hold the Confederates long enough for Meade to assemble his army, but also to have choices of action; he was not obliged to fight at Gettysburg and might have chosen to retire to another location.

Lee found himself in a position where the arrival of the constituent parts of his army from different directions forced him onto exterior lines, but the same factors gave his enemy interior lines. This might not have been a disaster if Lee had had enough troops, but he was outnumbered by the mid-point of the battle.

Lee put too much faith in the undoubted courage and high morale of his troops and asked too much of them. Meade – though very new to command of the whole army – had a realistic view of his forces and acted accordingly.

Lee was badly let down by Stewart. The lack of intelligence about the enemy's strength and movements helped to trap Lee into a battle that he did not want, or at least a battle in a place where he did not want one.

PART 4

THE RULES OF THE GAME

25. Some Tricky Questions

The proverb 'all's fair in love and war' is well known; likewise 'there are no rules in war'. If there is such a thing as 'all's fair' there must, presumably, be a concept of 'unfair'. There have certainly been attempts to form a basic code of law for conflict, though the extent of success is, to say the least, questionable. A good deal of any such code is clearly to do with humanitarianism, which does seem a little redundant in a business which is, according to many people, all about killing. In fact, war really is not 'all about killing'; on a strategic or tactical level it is all about attaining objectives and on a personal level it is all about not getting killed. The death of one or more enemy soldiers is not really relevant save that it may be – almost inevitably will be – impossible to prevent the frustration of our own plans without negating the enemy's willingness or capacity to fight, and that will generally mean that people will get killed. The objectives may be territorial, political, cultural or religious, but in most cases the aggressor would be content to achieve the objectives without killing anybody at all, and similarly the defender would mostly be happy to prevent the aggressor's success without loss of life.

The humanitarian aspect is not, however, unimportant and it is not – as a rule – a sort of moral fig leaf. Most combatants avoid the wilful killing of civilians because the majority of soldiers, like the majority of everybody else, just do not want to do that. Their humanity gets in the way of wanton murder – and a good thing too, of course. Indiscriminate killing is not a thing that most people are comfortable with, nor is it easy for most people to simply walk past an injured person who is clearly in need of help, but there is a host of factors that impinge on human feeling on the battlefield.

Stopping to tend to a wounded comrade is a natural action, but may be extremely undesirable from a military perspective. Quite apart from the question of whether the soldier can actually do anything to aid the wounded man, there is the matter of why he is on the battlefield in the first place. If he and his fellow soldiers stop what they are doing to tend to every wounded man, the attack or defence role that they are carrying out is going to be impaired at best and negated at worst. If the attack or defence fails completely, would the injured man be better served than if it had been successful? If the objective is not attained, how many more men will be wounded or killed in another attempt, and if a position is not retained how many more will die in an effort to recover it? Moreover, if the defence fails, will the enemy tend to the wounded man as well as your own medics? Will they tend to him at all or will they 'finish him off'? If he becomes a prisoner, will he be treated in a decent and humane manner? The latter questions may be of paramount importance to the wounded soldier if the general practice of the enemy is not in accordance with what we tend to think of as the 'rules of war'. Aiding a wounded enemy is a noble thing, and there are countless instances of quite remarkable heroism to be found, but it is not always the case. Throughout history, vast numbers of wounded soldiers who posed no threat whatsoever have been killed or just left to die of wounds or exposure.

We might make the same point about prisoners generally. Although it is 'right' to treat prisoners in a humane fashion it has never been universal practice. In fact, soldiers are not really required to take prisoners at all – though intelligence staff would very much prefer that they did, if at all possible. There may be good – or at least rational – reasons for this. There may be no means of securing prisoners or men to guard them; it may be impossible to take into custody those enemy soldiers who have lost the will to fight, and there are many examples of such men just being ignored in the hope that they will find their own road away from the fighting. Arguably, of course, such men would not be prisoners, but deserters if we accept the premise that a deserter is not simply a person who has wilfully absconded from the army, but one who has become separated from his unit and made no attempt to rejoin it.

It is not always physically possible to take prisoners. During the Falklands War, there was some criticism that the British submarine HMS *Conqueror*, having sunk the Argentinean cruiser *Belgrano*, did not make efforts to pick up the survivors. Any attempt to do so would have put *Conqueror* at risk from the two Argentinean destroyers (*Belgrano*'s escorts), which were actively pursuing her, but even if it had been possible, what was the commander of *Conqueror* going to do with 700 or more prisoners?

Surrendering is often a difficult business. A soldier can throw his rifle away and approach the enemy with his hands up, but offering to surrender is not the same thing as having that surrender accepted. Even where the general policy is to accept surrenders, a great deal will depend on the immediate circumstances. If there is a lot of firing going on it may be extremely difficult to tell whether someone is trying to surrender at all. A lot also depends on both the general cultural ethos of the enemy and how they see their opponents or – and this may be crucial – to the moods of the moment. Men who are (or have just been) getting shot at or shelled and have sustained casualties themselves are less likely to be amenable to accepting responsibility for people from the other side.

Surrender is not a guarantee of safety or of reasonable treatment. As long as the prisoner is in the front-line positions of the men to whom he has surrendered he is just as vulnerable as they are to incoming fire. Even if he is treated fairly by the men in the front line, he may suffer at the hands of others as he is passed along to a prison camp.

Of course, the individual may not have made the decision to surrender, but be part of a command which has surrendered en

Chinese captured by USMC in Korea. Now what do we do with them?

masse on the orders of its commander. Although there are 'rules' to cover the treatment of prisoners there is no surety that they will be observed, even by countries that are signatories to the various conventions that should apply. The use of prisoners of war as labour is more than just an ancient tradition dating back to days when prisoners of war became the slaves of the victors; it is enshrined in the Hague protocols and Geneva conventions and UN practices. Prisoners may be put to work, but not in hazardous conditions and not in war work, but who is to decide what is hazardous and what work is there in wartime that is not, in some way, a contribution to the war effort? Different governments and military establishments have had widely differing views on this subject. Axis POWs in Britain and Canada and Allied prisoners in Germany during the Second World War were often put to agricultural work, though clearly this was war work of a sort. It would be splitting hairs to claim that prisoners in Japan who were sent to mine coal or copper were doing 'war work' but that their counterparts in Canada were not, though clearly the treatment of prisoners in Japanese captivity was much closer – and morally identical – to forced labour in Nazi concentration camps.

Treatment of civilians is also the subject of 'rules' for the conduct of war, and is another nest of difficulties. Millions of civilians have died as an immediate consequence of combat without anybody actually targeting them. It would be easy to assume that the soldier should never fire his weapon at a location where there are – or might be – non-combatants, but that is easier said than done and, in fact, is not really a viable concept at all. Put yourself in the position of the man on the battlefield; if enemy troops are firing at you from a rooftop you will naturally fire back. You may call in an airstrike or artillery and demolish the building. If it transpires later that the cellars of the building were full of civilians seeking shelter from the battle, does that make you responsible for their deaths? If – and this is not unknown – the enemy has chosen to fire from a position to which he has brought civilians as a human shield specifically to deter you from firing, he is clearly responsible for putting them in harm's way. If you choose not to fire, the enemy has obviously gained a tremendous advantage, but are you then responsible for injuries and deaths sustained by your own force owing to to fire from that location?

Naturally, we would all like to think that soldiers will take reasonable care to prevent civilian casualties, but it is very difficult to ascertain what constitutes either 'reasonable' or 'care'. The deliberate or random shooting of civilians is obviously not acceptable, but it may be very difficult – or just impossible – to ascertain whether a death is an incidental consequence of being in a combat zone or murder.

There is also the thorny question of what constitutes a civilian and what constitutes a soldier. Preventing clear identification of combatant or non-combatant is an essential part of the armoury of every insurgency. Someone taking potshots at soldiers does not stop being a combatant because they lay their rifle aside, and we cannot expect soldiers to only select targets in uniform if the enemy does not wear a uniform. Fighting against a partisan (or guerrilla/terrorist/freedom fighter/bandit/insurgent – it is very much a cultural distinction, after all) enemy is fraught with difficulty for the soldier. His uniform identifies him as a participant, but his enemy is often only identifiable by the fact that he is carrying a weapon. It is general practice to recover the weapon of a casualty, if possible, but arms are generally in short supply for an insurgent force, so when a fighter is killed, it is that much more important to secure his rifle. A soldier shoots and kills an insurgent because he sees a man carrying a rifle. The weapon is picked up by another insurgent who makes his escape; the soldier is now at risk of being accused of the indiscriminate murder of a civilian simply because the evidence of military activity (the rifle) has been removed from the scene.

Making the combatant look like an innocent bystander is a common practice among insurgent forces, particularly in urban areas where there is more chance of getting some mileage out of the incident. If the insurgents have the support of the people, there is a good chance that (genuine) civilian witnesses will be prepared to say that the dead person was unarmed and/or that people who did not actually see the incident themselves will be prepared to tell the same story. Best of all, there might be a news team with cameras who will broadcast footage of the tearful witnesses right around the world. It can be genuinely difficult to distinguish between fighter and bystander, but there are other cultural complications. The conduct of war is largely seen as the province of men who, willingly or otherwise, have become soldiers. When a soldier shoots a woman or a child, there is a general tendency to assume that he has committed a murder, but this is not necessarily the case.

Over the last 100 years, the role of women in the military has developed from nursing and other ancillary functions. In the last few decades it has become increasingly normal for women to serve in regular armies in combat roles – we are no longer surprised that a woman should be a fighter pilot or the captain of a warship. The cultural assumption that women should not be deliberately selected as targets in war is no longer rational. If pilots have the opportunity to shoot down enemy warplanes, that is rather what we expect them to do. It is what they get paid for and it is the reason the government spent millions on their training and, of course, on the aircraft they are flying. The gender of the enemy pilot is really not an issue. Any suggestion otherwise is a redundant application of the playground adage that 'boys must not hit girls'. The bullet fired by a female soldier is just as lethal as one fired by a male soldier. The same applies to the tragic phenomenon of child soldiers. The age of the combatant is not a useful indication to his or her competence in battle; moreover, if a soldier is receiving fire, how is he –or she, of course – to gauge either the gender or the age of the assailant? More to the point, why should that be a factor in his or her decisions?

Of course, innocent bystanders do fall victim to being wounded or killed because combat happens to occur where they live or work, and it does happen that soldiers do kill civilians quite deliberately.

It can even be a matter of policy.

At its most extreme, attacks on civilians becomes genocide, which is invariably a political, ideological and/or religious imperative, not a military one, but it can have a military purpose. In 1296, Edward I's army destroyed the town of Berwick in Scotland and thousands of civilians were killed. The people posed no threat to Edward's army or his strategic intentions, but he was aware that Scotland was unaccustomed to war and that word of the events would spread quickly and lose nothing in the telling. Consequently, he encountered no further resistance from any Scottish town throughout the rest of the campaign. There is no reason to think that he was concerned that the resistance of any other town would inflict any serious damage on his force, but it might well slow his progress. His objective was conquest, not genocide, or even destruction; he did, after all, want to acquire Scotland as a profitable 'going concern', not a wasteland, but his actions at Berwick served his purpose very effectively, so the destruction of one town, however significant economically (and in 1296 Berwick was one of the most important towns in the British Isles), was a successful strategic move. Arguably, however, it was only successful because it was not repeated. If similar destruction had been adopted as a standard practice it might well have stiffened opposition elsewhere – there would be little point in surrendering a town intact if the people were going to be killed anyway.

Accepting the surrender of combatants, traditionally, imposes a degree of responsibility on the captor, but does the same apply to the people of a conquered and occupied territory? We would all like to think that an occupying power will behave in a civilised manner, and quite often that will be the case, though there will sometimes be a sharp disagreement about what exactly is a 'civilised manner'. There is a considerable body of opinion in Afghanistan that believes providing girls with the opportunity to go to school is an uncivilised, unjust and unreasonable imposition by foreign interventionists and that it is a duty to dissuade girls from going to school by the simple expedient of killing them.

The behaviour of an occupying power will often be influenced, if not dictated, by cultural or political factors, but it will also be affected by the behaviour of the local population and the nature and resources of the occupying power. Even if the general approach is benevolent in intention, it may well be quite impossible for the occupiers to provide the services that the people require. A consequence of war is often a shortage of food due to damage to crops and livestock or due to damage to the infrastructure, which prevents movement of goods or the functions of the marketplace. More often than not, these will be challenges beyond the capacity of the occupiers even if the will exists to solve the problems and, as always, the first responsibility of the commander on the ground is the preservation of his force. If there is a limited supply of food, the needs of his troops must have priority over those of the local population if he is not to face accusations of dereliction of duty.

It would be unfair to claim that ethics have no place in war, but the sad fact is that the pursuit of the crucial objective in war – victory – will tend to take precedence over all other considerations. It would certainly be preferable to have a strictly observed set of standards, but it would be impossible to get a universal agreement as to what those standards would be … and who would enforce them? International bodies can and do make a positive contribution, but they cannot observe every engagement all of the time and are always open to accusations of bias.

There is also the question of whether participants in a war accept that certain aspects of behaviour are necessarily wrong, or that this or that action is wrong for one side, but perfectly reasonable for the other. That is not quite as simple as it sounds. We are horrified if insurgents are executed out of hand once they have been captured. The 'laws of war' – such as they are – have been adapted to take account of combatants that do not have uniforms and who might otherwise be 'legitimately' shot as spies, but the converse gets very little consideration. A developed insurgency which has achieved political and military control over a portion of a country may be able to hold prisoners in a secure and humane environment, but a less developed one may not. During the Irish 'troubles' of the 1970s and beyond, it was assumed that the security forces of the UK government would take prisoners and keep them in confinement pending a political settlement, but the IRA (Irish Republican Army) and INLA (Irish National Liberation Army) could not have done such a thing even if they had wanted to. If soldiers or police officers were captured, where would the prisoners be kept? Who would guard them? What would the consequences be if they escaped? Would keeping them prisoner have a detrimental effect on the propaganda war? It is not necessary to approve of the execution of a prisoner to accept that there was no other practical course of action for the captor.

Afghan children. Write your own caption, draw your own conclusions; there is so much emotional 'content' that the shot looks posed, but it wasn't. (From Nick Allen's *Embed: To the End with the World's Armies in Afghanistan*, The History Press)

'Time is a jet plane, it moves too fast' (Bob Dylan). A matter of months before the outbreak of the Second World War, two veterans of Gettysburg drop flowers from the air onto the battlefield on Memorial Day. What kind of equipment will be available to combatants in twenty years, or even ten?

26. If You Get into a Fair Fight, You've Done Something Wrong

One person's dastardly attempt at cheating is another person's legitimate *ruse de guerre*, a brilliant and admirable tactical innovation rather than an underhand and unethical crime against military ethics or against humanity in general. Like the distinction between the bandit, the terrorist and the noble freedom fighter, the difference is largely one of perception or even wilful self-deception. This is a symptom of the individual perceptions of war in general as well as of specific conflicts in which we have taken a personal interest. We tend to acknowledge the validity of the proverb 'all's fair in love and war' except when it might impinge on the prospects of a cause for which we have sympathy or respect.

The mere fact that we can ever think in terms of 'fair' or 'unfair' in relation to warfare is an indication that we have deliberately chosen to disregard what should be a self-evident aspect of fighting. Combat is not 'fair', and governments, commanders and individual soldiers go to considerable lengths to ensure that it is not. They have every reason to do so. The purpose of war is not to conduct operations but to win. If victory requires a degree of moral vacuum, that is just part of the price of success.

It is not uncommon to compare war to games or sport. In a certain sense, this is not always wrong; chess and draughts – and of course war games on tabletops, maps or computers – are depictions of conflict to a greater or lesser degree, but the analogy is not that useful or desirable. Sport (or games) is about a 'level playing field' with rules and practices which affect both sides equally. War is just the opposite. The contenders want – or certainly should want – to gain every possible advantage. If the weight of advantage is so great that the enemy gives up, then so much the better.

Perceived similarities between war and sport are invariably quite preposterous. Some are offensive, such as the idea that killing some enemy soldiers after taking casualties somehow 'evens the score'. Some are simply naive, such as the idea that a cavalry charge was akin to a rugby scrum with men and beasts pressing through a body of enemy troops by the sheer weight of the rear ranks pushing the front rank onward. It sounds convincing to anyone who cannot ride a horse, but the reality would simply be a pile of horses which had tripped up over one another and which would not constitute a threat to the target.

New developments in warfare are often seen, or at least described (generally by the loser), as 'cheating'. The introduction of firearms was seen by some late medieval commentators as being 'wrong' or at least undesirable because it would allow a man of humble station to kill his social superiors without regard to the military/social ethos of the day. Similar comments were made in relation to archery and there is still a certain resonance, though not one that really bears much examination. English historical romance puts much stress on the 'lower classes' defeating French aristocracy with the longbow, but, in fact, the classic longbow battles were few and far between and a great many archers were not 'mere peasants' but men of a professional-class background. In fact, there are many examples of men whose social status would generally have put them in the gentry class serving as archers – partly, no doubt, because of much lower expense but also because it was 'cool' to be an archer. The high incidence of middle-class men in the volunteer forces or in the rifle clubs of the nineteenth century is not dissimilar in principle.

Social conservatism has continued to be a factor in modern times. Churchill was initially opposed to providing sub-machine guns for British soldiers on the grounds that such things were weapons for gangsters, but there was also a cultural throwback to an idealised view of marksmanship rooted in the tales of Robin Hood.

Most developments in the technology of war are viewed as 'barbarous' by at least one side of an argument, rather neatly sidestepping the point that the whole business is barbaric in nature, though that may in itself be a slur on the nature of barbarians (leaving aside the issue of what we mean by 'barbarians'), who all in all were inclined to fight for more tangible results than ideological or religious motives. The introduction of firearms, gas, mines, barbed wire, tanks and submarines have all been subject to the claim that one of the protagonists – inevitably 'them' rather than 'us' – are not playing the game by the rules. Similar points are made about guerrilla fighters; they won't 'stand up and fight', though obviously standing up in the open is hardly a way to conduct operations on the modern battlefield, nor indeed for a long time past. Even that distinction is defined by our point of view. Resistance fighters in occupied Europe in the Second World War are seen – and justifiably so – as brave men and women struggling against an oppressive regime rather than unauthorised combatants taking potshots and detonating improvised explosive devices from positions of relative safety.

Combatants must be prepared to use any and every tool at their disposal to obtain victory, which is the only real target in war. The purpose of war may be acquisition or to resist annexation, but the purpose of fighting is to win. There is, of course, a difference between being notionally prepared to use a certain weapon and actually doing so. Any number of political or cultural imperatives may prevent the deployment of a specific weapon or technique. One could make an argument that the Falklands War could have been brought to an end with much less loss of life by the threat of a nuclear weapon being dropped on Buenos Aires, but in practice it would have been politically unacceptable within the United Kingdom, the NATO Alliance and, indeed, the rest of the world, even though deterrence – which is simply the threat of massive and immediate harm on a gigantic scale – lies at the root of the theoretical defence ideology of Britain. Our fascination with the bomb has diminished somewhat, as we have come to believe that nobody – or at least not the established nuclear powers – is actually prepared to use it. In a sense, the opposition to nuclear weapons is something of a curiosity. If one is to suffer death in war it is hard to see what difference it makes whether we are killed by a hydrogen bomb or an arrow through the temple – we are still going to be dead.

Reluctance to use 'the bomb' is not simply a matter of political or diplomatic expediency; there is a moral dimension that can be found right across the political spectrum. Contrary to what we might expect, Ronald Reagan, generally seen as a 'cold war warrior', was steadfastly opposed to the existence of, never mind the actual use of, nuclear weapons, and was involved in anti-nuclear politics from their very inception, to the degree that he planned to appear at a nuclear disarmament event as early as December 1945. His support for the 'Star Wars' programme was aimed at making 'the bomb' obsolete and achieving a nuclear-free world; he even made it clear to his military staff and advisors that there were absolutely no circumstances in which he would authorise a nuclear strike. This was a brave and morally laudable stance to take, but it was also a matter of humane standards – which he saw as Christian values – and an acceptance of a practical truth. If millions were to be killed in an attack on the United States, how would it help anyone to cause millions more deaths in Russia or China? As with so many things – in fact, virtually everything – that we feel we 'know about war', the substance is far removed from the belief.

The 'Good War'.

Author of the most famous work on war theory, *The Art of War*. Are his observations still applicable to warfare today?

Appendix
Master Tzu and the Art of War

OR

'IF YOU'RE ONLY GOING TO READ
ONE MORE BOOK …'

There are a handful of names which are, or certainly should be, familiar to all students of warfare, first and foremost being Sun Tzu, or 'Master Tzu'. The title of this chapter can be construed in three ways, which is actually a very 'Sun Tzu' approach. We can read it as 'Master Tzu' – which is simply his name – or as Tzu, master of his subject, or as an injunction to read and learn; if we can 'master' Tzu then we can 'master' the art of war, or at least understand the principles which are the foundation of success in conflict.

Numerous scholars have cast some doubt on his very existence, in the sense of being one single individual who lived in China 2,500 years ago, and it is widely accepted that, assuming he did exist, his work was developed and extended by his descendent Sun Ping (his son according to some writers). It is certainly the case that several of the examples and precepts in his book *The Art of War* or *The Thirteen Chapters* begin with statements like 'Sun Tzu said …' which perhaps we should accept as an indication that the volume which we can consult today is not simply the work of one man.

Moreover, we should bear in mind the possibility, perhaps even the probability, that this remarkable (and comfortingly brief) volume did not spring from a single mind in the first place. If we accept the traditional account of Sun Tzu's life, we find him serving as the commander of the army of King Helu of Wu, the culmination of a career as a professional soldier. As such, we should naturally assume that he had garnered a good deal of personal professional experience, but it may well be the case that his work is not simply a reflection of his own experience and intellect, but also that of the officers with whom he served and – for all we know – owed a good deal to an existing body of written work which no longer exists and was not, perhaps, particularly widely known at the time. We can see a similar relationship between the work of Christine de Pisan, who virtually nobody has ever heard of, and that of Niccolo Machiavelli, who virtually everyone has heard of.

The origin of *The Thirteen Chapters* is not the issue here; the sheer brilliance is. The simplicity of his precepts can be misleading; a short paragraph demonstrates a principle which seems so straightforward it should be obvious to anyone, but there is invariably at least one deeper insight (and frequently several) if we care to look for it.

According to tradition, Sun Tzu had already finished writing his book when he obtained the post of chief general of the army of the kingdom of Wu by demonstrating his understanding of the principles of successful command. The story goes that Sun Tzu gathered the king's concubines and appointed two of them – the king's favourites, as it happened – to take charge of the rest and lead them through a series of military manoeuvres.

Unsurprisingly, the leaders were not especially impressed and made no real effort to understand what was required of them or to instil a sense of discipline into their charges. Consequently, when orders were given, the concubines laughed and larked around since they took neither their newly appointed leaders nor the task in hand seriously. Sun Tzu promptly had the leaders executed, which had a predictable effect on the rest of the group. The story may well be apocryphal, but it conveniently demonstrates a fundamental element of the Sun Tzu approach or philosophy. One short and simple example is used to demonstrate several different, though interconnected, aspects of the business of war.

We might take the moral of the tale to be the value of rigid discipline and even perhaps a suggestion that iron discipline is best enforced with iron methods, and effectiveness of command through discipline is one of the lessons, but it is crucial to our understanding of the Sun Tzu 'teaching methodology' to see that discipline is only one part of the lesson and that the other parts are of equal importance. The unfortunate leaders had not understood their duties or the purpose of the exercise and had, almost inevitably, failed to achieve their objective; if the commander does not understand his duties it is virtually inevitable that the troops will not be effective. They will not understand their roles, the purpose of their efforts, the value of acting in concert or the goals of either their training or its application in the field. The lesson has, therefore, a real value for both the leader and the led. There was also a lesson for the king himself. If he does not see to it that his officers have been selected wisely and are properly prepared to undertake their tasks, he is inviting defeat. He may have gathered the troops, he may have armed them to the very best standards, he may have ensured that they are well fed, well shod and have the best possible terms and conditions, but they will not be effective on the battlefield if they are not properly structured and trained *for the task in hand*. The formal nature of court life in classical China would suggest that the concubines would have had a clear understanding of the nature of a hierarchical structure and doubtless many of them would have skills beyond those that we might immediately expect – perhaps as musicians or artists – but the nature of their hierarchy was not well suited to forming an effective unit for battle. The unfortunate leaders doubtless enjoyed greater status, since they were the king's favourites, but they did not have the right sort of authority – or perhaps the personal ability – to demand respect and obedience. That would be another lesson for the king; appointing people to command positions on the basis of how much he liked them or on the basis of social rank was not the best means of selection. That lesson was hammered home through the loss of his two favourite concubines, which was in itself a demonstration of a principle; the consequences of poor selection would be painful.

In one short example, Sun Tzu had shown the importance of several key elements to success in war. The troops must be chosen from the most suitable portions of the community, the troops must be trained, and they must be led by people who understand what they are doing and why they are doing it. The leaders themselves must be selected on the basis of ability, not personal or social consideration, and they must be trained to a specific purpose, that failure to provide an appropriate structure with well-developed discipline is critical, and that the object of the exercise – in this case building the sort of effective unit that is vital to success – must be maintained at all times if the whole process is not to end in defeat.

The wider lesson, of course, was that all of these elements must be brought together as one; none of them has very much value in isolation. We might make a similar observation about one of Sun Tzu's better-known examples. In this tale, a commander is faced with the problem of dealing with an opponent who has a strong force, a strong position and a very good reputation as a tactician. He decides to goad his opponent by writing to him: 'If you are such a great general, why don't you come down to the plain and fight', which prompts a response of, 'If you're such a great general, why don't you *force* me to come down and fight'.

The first lesson is pretty clear-cut: never let the enemy goad you into an unwise manoeuvre, but there are also what we might think of as sub-textual implications. The invitation to 'come down and fight' shows that the general 'on the plain' has already decided that the position held is too strong for a successful attack; he has, in essence, given up on the current situation. Clearly, that is an admission of weakness and an indication that the defending commander on the high ground is safe for the foreseeable future and that his opponent has run out of ideas, but cannot accept a stand-off. The attacking commander may, therefore, commit himself to a less-than-ideal course of action simply because he has no better option. If he chooses a poor approach he will very likely provide his opponent with an opportunity which would not otherwise have arisen. By not accepting the challenge, the defending general is making life harder for his opponent, which is, in itself, a desirable thing in war, but he is also keeping his own options open, which is all to the good.

Even if the general down on the plain chooses to make no approach, his overall position will suffer. His troops will become bored, apathetic and demoralised – which may, in due course, make his force vulnerable to a surprise attack – but they will also be a drain on the resources of the government. The army is in the field and in the face of the enemy, so potentially (if not inevitably) the costs will be astronomical, but nothing is being achieved. This, in turn, will eventually have an impact on the political, social and economic structures 'back home'. Any domestic political opposition to the war as a whole, or even to just that particular theatre or campaign, is likely to be strengthened by the lack of results, and the longer the troops are in the field and away from their families and their jobs the greater the social and economic damage.

To some extent this would also apply to the defending army. They too are in the field and costing a fortune to no immediate effect and they too will eventually become demoralised through inaction, but the military environment is more favourable to them than the army on the plain. They know (or at least it is the commander's duty to ensure that they know) that they occupy the better position, that the enemy does not dare to attack them, so they can deduce that the opposing commander is less talented than their own. If that were not the case then they would have been manoeuvred out of their position already. There is also an aspect of 'opportunity cost'. The defending commander and his army are 'doing their job' simply by staying put; their purpose is to defend a location. The attacking force is not only failing to achieve anything in that location, they are being prevented from conducting operations anywhere else.

Once again, there is another lesson implicit in Sun Tzu's 'parable'. It is possible to achieve victory in war without bringing about a cataclysmic battle. As a general rule, a war will involve fighting, but in this example there is an implication that the attacking general may have to abandon his campaign without a resolution through combat. An army cannot be kept in the field forever; even if the economic strength and the political will exist, eventually the price paid in desertion, demoralisation, disease and other factors will force the careful commander to withdraw.

We can see an example of this in the Weardale campaign of 1327. Edward III brought a large and well-structured army to the field, but found himself constantly outmanoeuvred by the Earl of Moray. Invitations to 'come out and fight' were met by a response of 'make me if you can'. In due course, the English army ran out of supplies, became demoralised by regular defeats in small skirmishes. It was only by sheer good fortune that Edward himself was not captured in a night raid. As the days passed, men deserted in droves and in due course the campaign was abandoned. The cost in both political and economic capital was enormous, so Edward was unable to mount a fresh expedition and was forced to negotiate a peace settlement that recognised Scottish independence and the legitimate kingship of Robert Bruce. Interestingly, Robert had not been able to achieve that outcome through a stunning victory at Bannockburn in 1314. In a political sense – and all war is political in nature – more had been attained through avoiding battle than through winning one; a clear validation of the Sun Tzu approach to war. It may be necessary to fight, but it is not always either desirable or effective to do so.

Another of the more popular propositions from Sun Tzu is the advice to 'know your enemy as you know yourself and you will not be defeated in a hundred battles'. Several concepts are contained in that single sentence, and it would be all too easy to see it as a rather obvious statement; but Sun Tzu (as ever) wants the reader to really concentrate on the meaning, not glide over it as the cliché it has become. To 'know your enemy' is not simply to recognise his existence, to evaluate his numbers and weapons or the commander as an individual. Of course, 'knowing the enemy' encompasses these things, but they are only three factors among many. Numbers and armament are certainly important, but how does the enemy usually apply them? What other options has he used in the past? What new applications might there be in the future? If the enemy has had success in recent times, will he continue with tried and tested methodology or will he try something new in the belief that his opponents may have learned from their own defeat? Sun Tzu is not simply advising the student to be aware of the enemy army as an institution, but to continually question what developments may arise in the future. In the same breath, he is asking the student to apply those questions to his own force. It is not just a matter of knowing what the force can do at the present time or what it has been able to do in the past, but of what it might achieve by alternative approaches in the future.

It should be obvious to us all that understanding the enemy and his force is critical to the application of our own, not simply in the sense of what we can do, but also in being realistic about what is beyond our capacities. The latter part of the maxim is more subtle than it first seems. The implication is not that you can achieve victory over an enemy one hundred times (he means 100 per cent of the time) because you understand his force and your own, but that knowledge will enable you to make valid judgements about when to fight and when to avoid battle. In essence, he is saying that if your intelligence is absolutely sound and if there is a clear advantage to be gained, and if the overall military environment – terrain, weather and the wider tactical, strategic and political factors – are all favourable, you should be prepared to offer, accept or force battle with confidence of success. If they are not, you should avoid battle if at all possible.

The problem, of course, is that the commander cannot know absolutely everything about himself, let alone a military force under his command, and he certainly cannot know absolutely everything about the infinite range of factors which might affect the outcome of fighting a battle – or of not fighting one. All he can do is equip himself the best he can and hope that the imponderable elements (the 'friction' described by Clausewitz) will not work more heavily against him than against his enemy. Sun Tzu understood this and tells the reader that battle is always a gamble, a matter of matching resources to risk, risk to opportunity, opportunity to reward. The subtitle of this chapter is 'If you're only going to read one more book …'. The answer to the implicit instruction is … make it *The Art of War*. It is not 'hard reading', but it has to be 'read hard' (and probably several times) if the reader is to gain the greater benefit, but the rewards are invaluable for the reader who is willing to make the effort.

Further Reading

The following handful of volumes is not intended as a comprehensive reading list, just a selection of books which may help to carry the student a little further toward an understanding of conflict, to the level that one might expect for an undergraduate course in War Theory as opposed to War Studies, though the two should, in my view, be regarded as complementary and overlapping disciplines. Some of these books are very 'readable', others not so much so, but perseverance is recommended!

I have deliberately avoided volumes on the operational doctrines of different armies, the conduct of particular conflicts or the military careers of individual officers – many of which have considerable value – because the topic of this book is not a war, or even a period of wars, but of war as a subject in its own right.

Alfred Burne, *The Art of War on Land*.
Archer Jones, *The Art of War in the Western World*.
Carl von Clausewitz, *On War*.
Christine de Pisan, *The Book of Feats of Arms and Chivalry*.
Dandridge Malone, *Small Unit Management*.
Greg Cashman, *What Causes War*.
Hew Strachan, *European Armies and the Conduct of War*.
Honoré Bonet, *The Tree of Battles*.
James Dunnigan, *How to Make War*.
John Keegan, *The Face of Battle*.
Mao Zedong, *On Guerrilla Warfare*.
Michael Handel, *Masters of War*.
Michael Howard, *Empires, Nations and Wars*.
Niccolo Machiavelli, *The Prince*.

S.L.A. Marshall, *Men Against Fire*.
Simon Trew and Gary Sheffield (eds) *100 years of Conflict, 1900–2000*.
Sun Tsu, *Art of War*.
USMC, *Warfighting*.
Trevor Dupuy, *Numbers, Prediction and War*.
Wesley Clark, *Winning Modern Wars*.
William Seymour, *Yours to Reason Why*.

Index

If you enjoyed this book, you may also be interested in…

Who Takes Britain To War?

JAMES GRAY MP & MARK LOMAS QC

978 0 7509 6182 0

The long-standing parliamentary convention known as the 'Royal Prerogative' has always allowed Prime Ministers to take the country to war without any formal approval by Parliament. The dramatic vote against any military strike on Syria on 29 August 2013 blew that convention wide open, and risks hampering Great Britain's role as a force for good in the world in the future. Will MPs ever vote for war? Perhaps not – and this book proposes a radical solution to the resulting national emasculation. By writing the theory of a Just War (its causes, conduct and ending) into law, Parliament would allow the Prime Minister to act without hindrance, thanks not to a Royal Prerogative, but to a parliamentary one.

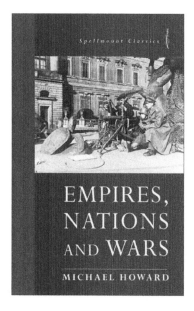

Empires, Nations and Wars

MICHAEL HOWARD

978 0 86227 372 6

Empires, Nations and Wars brings together the articles and lectures of Sir Michael Howard during his time as Regius Professor of Modern History in the University of Oxford between 1980 and 1989. Some of the articles in this work reflect on contemporary events, but most are concerned with the historical process which underlies international politics.

Visit our website and discover thousands of other History Press books.

www.thehistorypress.co.uk

Lightning Source UK Ltd.
Milton Keynes UK
UKOW07f0609301114

242395UK00002B/2/P

9 780750 959728